Earth, Cosmos and Culture

This book traces the development of diverse British cultures of outer space, utilizing key geographical concepts such as landscape, place, and national identity.

It examines the early visionary ideas of writers H G Wells and Olaf Stapledon, the ambitious British space programme of the 1960s, and narrations of British cultural identity that accompanied the space missions of Helen Sharman, Beagle 2 and Tim Peake. The exploration of British cultures of outer space throughout the book helps to understand the emergence of the British Interplanetary Society. It also explains its significance in the pre-war and post-war periods through an analysis of the roles of influential figures such as Arthur C. Clarke and Patrick Moore. The chapters explore utopian and dystopian representations of space exploration, examine the mysterious phenomenon of UFO culture, and consider plans for humanity's imagined future across interstellar space. Throughout the book, geography is advocated as a home for critical studies of outer space, illuminating its significance in terms of the reciprocal relationships between exploration and the sublime, science and the imagination, Earth and cosmos.

As an emergent field of research in the social sciences, this book makes an excellent contribution to the cross-disciplinary study of the outer space in Britain and abroad developing a distinctive kind of outer spatial geography with major implications for future teaching and research.

Oliver Tristan Dunnett is a lecturer in human geography at Queen's University Belfast. His research focusses on the ways in which the cultures and politics of outer space, science and technology are connected to questions of place, landscape and identity in a variety of local, regional and national contexts.

Routledge Research in Historical Geography

This series offers a forum for original and innovative research, exploring a wide range of topics encompassed by the sub-discipline of historical geography and cognate fields in the humanities and social sciences. Titles within the series adopt a global geographical scope and historical studies of geographical issues that are grounded in detailed inquiries of primary source materials. The series also supports historiographical and theoretical overviews, and edited collections of essays on historical-geographical themes. This series is aimed at upper-level undergraduates, research students and academics.

Architectures of Hurry
Mobilities, Cities and Modernity
Edited by Phillip Gordon Mackintosh, Richard Dennis, and Deryck W. Holdsworth

Anarchy and Geography
Reclus and Kropotkin in the UK
Federico Ferretti

Twentieth Century Land Settlement Schemes
Edited by Roy Jones and Alexandre M. A. Diniz

Resisting the Rule of Law in Nineteenth-Century Ceylon
Colonialism and the Negotiation of Bureaucratic Boundaries
James S. Duncan

Cold War Cities
Politics, Culture and Atomic Urbanism, 1945–1965
Edited by Richard Brook, Martin Dodge and Jonathan Hogg

Micro-geographies of the Western City, c.1750–1900
Edited by Alida Clemente, Dag Lindström and Jon Stobart

Earth, Cosmos and Culture
Geographies of Outer Space in Britain, 1900–2020
Oliver Tristan Dunnett

Earth, Cosmos and Culture

Geographies of Outer Space in Britain, 1900–2020

Oliver Tristan Dunnett

Routledge
Taylor & Francis Group

LONDON AND NEW YORK

First published 2021
by Routledge
2 Park Square, Milton Park, Abingdon, Oxon OX14 4RN

and by Routledge
52 Vanderbilt Avenue, New York, NY 10017

*Routledge is an imprint of the Taylor & Francis Group,
an informa business*

© 2021 Oliver Tristan Dunnett

British Library Cataloguing-in-Publication Data
A catalogue record for this book is available from the British Library

Library of Congress Cataloging-in-Publication Data
A catalog record has been requested for this book

ISBN: 978-0-8153-5628-8 (hbk)
ISBN: 978-0-367-76240-7 (pbk)
ISBN: 978-0-8153-5630-1 (ebk)

Typeset in Times
by KnowledgeWorks Global Ltd.

Contents

List of Figures   vi
Acknowledgements   viii

1 Introduction: Geographies of outer space in Britain   1

2 Science-fictional foundations: A comparative literary
   geography of H G Wells and Olaf Stapledon   12

3 Synthesising outer space: The British
   Interplanetary Society   36

4 Outer space and popular culture in post-war Britain   61

5 The British space programme: Geopolitics and empire   88

6 Interstellar exploration: Project Daedalus and
   the extra-solar universe   111

7 Space exploration, science and nationalism   135

8 Conclusion: Diverse cultures, possible futures   161

Bibliography   178
Index   188

Figures

2.1 Illustration by Henrique Alvim Corrêa, from the 1906
Belgium (French language) edition of The War of the Worlds.
The original caption stated: 'Martian Fighting Machine in
the Thames Valley'
Source Credit: Henrique Alvim Corrêa/Public domain   28
3.1 First London meeting of the BIS in October 1936
Source Credit: British Interplanetary Society   43
3.2 Front cover of the BIS Journal displaying diagrams of
the BIS space-ship
Source Credit: British Interplanetary Society   53
4.1 Nigel Kneale on the set of Quatermass II, 1955
Source Credit: BBC Photo Library   63
4.2 Patrick Moore pointing out the position of Capella
on *The Sky at Night*, 1960
Source Credit: BBC Photo Library   76
5.1 Front cover of Society for Space Research (*GfW*) pamphlet
for the 1951 International Astronautical Congress
Source Credit: British Interplanetary Society   95
5.2 Last page of Society for Space Research (*GfW*) pamphlet
for the 1951 International Astronautical Congress. The captions
translate as 'The London Congress'/'will they all come under
one hat?'
Source Credit: British Interplanetary Society   96
5.3 Map of Australia showing the Woomera Rocket Range
Source Credit: Commonwealth of Australia Department
of Defence   100
6.1 Illustration by Bill Dillon showing the construction
of the Daedalus vehicle
Source Credit: Bill Dillon/British Interplanetary Society   122
6.2 Daedalus model (second engine stage) by Terry Regan,
commissioned by the BIS in 2011
Source Credit: British Interplanetary Society   130

7.1 Crew photo for Soyuz TM-12, featuring cosmonauts
     Artsebarsky, Sharman and Kirikalyov
     Source Credit: spacefacts.de/Joachim Becker   142
7.2 Helen Sharman's launch couch and spacesuit, exhibited
     at the National Space Centre, Leicester
     Source Credit: National Space Centre   144
7.3 Andrew Birch 'Young British Artists' cartoon from 2002
     featuring the Beagle 2 lander
     Source Credit: *Private Eye* and Andrew Birch   149
7.4 Tim Peake at the launch of the UK Space Agency
     in April 2010
     Source Credit: UK Space Agency   151
8.1 LaunchUK promotional image
     Source Credit: UK Space Agency   166
8.2 Katie Paterson, All the Dead Stars, 2009
     Photo © Mead Gallery installation view, Mead Gallery,
     Warwick Arts Centre, 2013
     Source Credit: Katie Paterson Studio   171

Acknowledgements

This book has been in the making for over ten years, and a multitude of people and organisations have provided academic encouragement, facilitated the research and writing process and offered moral support. At the University of Nottingham, the supervision of David Matless and Mike Heffernan was inspirational in terms of providing a license to consider outer space as a subject of enquiry in cultural and historical geography, in a scholarly environment that provided a rich introduction to cultures of geographical enquiry. I am also grateful to the School of Geography at Nottingham for awarding me a studentship for the duration of my doctoral studies, without which this project would not have come to fruition. At Queen's University Belfast, colleagues have been incredibly supportive of this work, in facilitating sabbatical leave, in reading various draft chapters, and in providing company during countless coffee-breaks. Particular thanks to Diarmid Finnegan, Nuala Johnson, David Livingstone, Merav Amir and Tristan Sturm for this. I am also indebted to the wider academic and student community in geography, which has provided invaluable feedback on aspects of this research, whether through referees' reports for academic journals, questions and discussion at seminars and conferences, or indeed through a generally supportive and collegiate atmosphere. The three anonymous reviewers of the draft manuscript of this book offered detailed, constructive, and exceedingly helpful comments, for which I am very grateful. Any omissions, mistakes, misunderstandings or other errors are, however, entirely my own. Thanks also to all those who, working in libraries, archives and other research organisations, have kindly allowed access to materials and workspace to investigate key source materials, and for providing permission to reproduce certain images and text. I am grateful as well to all those at Routledge who have helped in the production of this book and for offering me the chance to write it. On a personal level, my heartfelt thanks to friends and family for their continued support and encouragement, especially to Felicity and Catriona.

1 Introduction

Geographies of outer space in Britain

In my time, I've been very fortunate to see many of my dreams come true. Growing up in the 1920s and 1930s, I never expected to see so much happen in the span of a few decades. We 'space cadets' of the British Interplanetary Society spent all our spare time discussing space travel – but we didn't imagine that it lay in our own near future.

- Arthur C Clarke, in his 90th birthday reflections

Cosmos is a thing of formal order and beauty.

- Denis Cosgrove, describing the work of Alexander von Humboldt

This is a book about outer space, its relation to British culture and politics since the start of the twentieth century, and to geographical enquiry. It originated from a long-standing fascination with spaces of the imagination, science and technology, provoked by many of the iconic images of space exploration that proliferated Western culture throughout the twentieth century. This is a fascination that has by no means been restricted to any one time or place, as dreams of outer space are likely to be as old as humanity itself. Yet such imaginations, no matter how expansive, are inflected by a sense of place and associated cultures and politics, and this book aims to understand how outer space has been understood in the specific context of the United Kingdom, from the start of the twentieth century to the present day, using contemporary geographical approaches.

As the writer Arthur C Clarke pointed out in his 90th birthday reflections, uploaded to the internet in the early years of online video-sharing, the twentieth century was a period that witnessed rapid technological change, and in no area was this more apparent than in space exploration.[1] Clarke had a significant influence on British understandings of outer space, perhaps more than any other individual, and his life-long role as someone who engaged with both the realm of the imagination and the world of science and technology reflects the outlook of this book, which is to say that, in order to fully understand the meanings of space exploration, it is essential to seek out the cultural and political roots of its scientific and technological discourses. As such, Clarke's involvement with the British Interplanetary Society (BIS), his early science-fiction novels, as well as some of his technical concepts of space exploration, are featured in a number of the forthcoming chapters, as prime examples of the cross-fertilisation of science and culture in under-standings of outer space in Britain. In terms of organisations, the BIS was the most influential group in conceptualising outer space in Britain, whose significance stemmed from its straddling of technological and imaginative themes, and its continued longevity as the oldest space advocacy group in the world. While this book by no means offers a comprehensive account of this organisation, various episodes in its history are examined in several chapters, including its establishment in 1930s' Liverpool, its post-war plans for configuring space exploration, and its expansive designs for interstellar flight. As such, the BIS encapsulates the dual nature of space exploration, as being part imagination, part techno-science, and in focussing on the cultural and political roots of space exploration in Britain, this book aims to broaden the scope of historical studies of outer space to take account of this hybrid quality. Indeed, when considering states such as the United Kingdom, in contrast to the principal space powers of the USA and Russia, acknowledgement of the imaginative as well as the technical becomes all the more important in appreciating the fundamental meanings of human spaceflight, space exploration and space science, and our complex relation to outer space here on Earth.

In such ways, this book argues that Britain became a home to rich dis-courses of outer space, both feeding from and contributing to iconic achieve-ments in space exploration, while also embracing the cosmos in imaginative and philosophical ways.[2] Cognisant of this spatial context, a central aim is to demonstrate how contemporary geographical enquiry can provide spe-cific and valuable perspectives from which to understand outer space. This is an argument that was initiated by Denis Cosgrove, and his critique of Alexander von Humboldt's seminal work *Cosmos* helped to demonstrate geography's special relevance to thinking about outer space.[3] The key the-matic areas which provide the interface for this book's research, therefore, are the cultural, political and scientific understandings of outer space; the context of the United Kingdom since the start of the last century; and the geographical underpinnings of their relationship.

Contextualising geographies of outer space in Britain

Outer space has long provoked fascination in the human mind, examples of which can be traced back to Classical antiquity with whimsical stories of transcendence into the heavenly realm, such as Lucian's *Icaro-Menippus* of the second century, to the early ages of the Enlightenment in Western cul-ture, in the revolutionary observations of Galileo and the imagined cosmic perspectives of Johannes Kepler, and later in the emergence of astrophysics, as the mysteries of the Universe have become deeper and more complex than had ever before been imagined. Yet the twentieth century was a time in which a profound shift in humanity's relation to the cosmos took place – the advent of spaceflight with the satellite *Sputnik* in 1957 and the human transcendence from Earth's gravity that was represented by the first cosmonauts and astronauts of the early 1960s. These, of course, were far from isolated events, and the cultural and political contexts of the 'space race' between the world superpowers of the twentieth century have been well-documented.[4] Recently, these contexts have been shown to extend far beyond the immediate events of the Cold War or even the rocketry programmes of the Second World War that accelerated the onset of spaceflight. Indeed, researchers have shown how the roots of spaceflight can be traced to earlier origins, including early-twentieth-century scientific networks in Imperial Russia or the 1930s' experiments of rocketry pioneers in California.[5] The importance of cultures and politics of outer space outside of the bipolar Cold War nexus has also been recognised, with 'European astroculture' being shown to have had rich and diverse meanings that have often been overlooked in studies of outer space, while the colonial roots of European space exploration have also been acknowledged.[6] What sets twentieth-century studies of outer space apart from studies in earlier periods is, therefore, the prospect of spaceflight and the associated cultures and politics of space exploration. While themes of astronomical observation or heavenly transcendence still come into play in this period, what tie together this book's focus on the twentieth century are the enduring possibilities of spaceflight, including its anticipation as well as its ongoing legacy. Studying fully the cultural and political contexts of spaceflight, then, involves considering a wide range of scientific, technological and imaginative sources that reach beyond the immediate environs of contemporary space agencies and individual heroes of the Space Age, to incorporate places whose connections to narratives of space exploration have been left relatively unstudied.

The United Kingdom has been understood as a relatively minor player in space exploration, with a paucity of human spaceflight experience in comparison to European and other international states, partly compensated by an established profile of private-sector space manufacturing and service companies. Yet there is a more complex story to tell about the UK's past and continuing role in outer space that considers the Cold-War-era efforts to produce a British-made satellite, the establishment of a British space-rocket to be launched from an anticipated spaceport in Australia, and the promotion of British private companies' involvement in the emerging space sector. There was, indeed, a period following the Second World War, amid an international nuclear arms race, in which the UK was briefly a world leader in outer-space research and rocket technologies.[7] Looking back further into the pre-war period, when workable concepts of space exploration started to emerge for the first time, British groups were among the first to realise the technological potential of rocketry for reaching outer space, as evidenced by the recollections of Arthur C Clarke and his associates in the British Interplanetary Society. Furthermore, when considering the science and technology of outer space together with the broader cultures and politics of space exploration, including such registers as science fiction, astronomy and folklore, deeper connections can be identified. In this way, an approach that looks beyond national programmes and frontline geopolitical rivalries has a lot to offer in studies of outer space. Furthermore, such accounts help to understand the changing fortunes of British society from the start of the twentieth century, including the ways in which British geopolitical outlooks altered from a position of global dominance to becoming increasingly reliant on a range of international alliances, as well as the shifting status of British science and technology and the effects these changes had on the lives of people in the UK. Taken together, considerations of British scientific and technological achievement in outer space, couched in their own cultural and political contexts, alongside the legacy of imaginative representations of outer space, makes possible a rich narrative of engagement with outer space in the modern era. Indeed, a key argument of this book is that the UK had an active outer-space culture during and beyond the twentieth century, as much as any other nation of the world.

Considering outer space in the UK context from the start of the twentieth century raises questions about how British cultures and politics of outer space have differed from those elsewhere. Indeed, this becomes one aspect of the 'geographies of outer space' that this book deals with, thinking about the ways in which particular spatial contexts, places and landscapes have influenced understandings of outer space, both official and popular. It is further argued that the diversity of approaches within the broad discipline of geography offers the potential for multiple analytical perspectives from cultural, political, historical and environmental geographies, across a wide range of possible case studies.[8] Over and above such approaches, however, there is a more fundamental reason why geographical enquiry offers particular advantages and proclivities to studying outer space, as iterated by Denis Cosgrove and others.[9] When Cosgrove wrote about Humboldt's nineteenth-century treatise *Cosmos*, that incorporated sections on the stars and planets as well as the features of the Earth, he was highlighting a long-standing tradition in geographical scholarship in which the spheres of the Heavens and the Earth were regarded in unison, as part of a divine creation, whereby one could not properly be understood without the other. Often represented in the twin production of terrestrial and celestial globes, this theme can be traced throughout Western culture, from the thinkers of the Classical period through to the Renaissance, including figures such as Aristotle, Ptolemy and Abraham Ortelius, and is a duality that, while diminished, can still be recognised today in some common atlases.[10] Yet for Cosgrove, Humboldt's work, in its structure, tone, and modes of representation, signalled a shift in geographical thinking away from this historic tradition of cosmography, towards understanding the Earth as a two-dimensional realm of surfaces and territories, jettisoning the cosmic entirely from the professionalised discipline of geography. This was a move that was caught up in the imperatives of European colonialism, mapping, and the advent of geopolitics, the vestiges of which endured well into the twentieth century in a range of sub-disciplinary traditions. However, human geography's subsequent embracing of post-modernist philosophies, including the cultural turn in geography and a geographical turn in the social sciences, for Cosgrove, offers interesting opportunities for re-connecting with the cosmic spheres of old. The onset of space exploration has further provided an opportunity for this re-appraisal of humanity's connection to the spaces beyond the Earth's surface, supported thematically by new geographies of aerial, subterranean and other three-dimensional spaces.[11] In looking above and beyond the Earth's surfaces, while considering the representation and experience of outer space as intimately connected to Earthly spaces, this book aims to re-kindle these deeper traditions of geographical enquiry, fostering a new geography of outer space that holds the cosmic and the terrestrial together in alignment.

Methods and sources

This book takes a selective sample of case studies to illustrate some of the predominant British cultures and politics of outer space since the start of the twentieth century. Examples were selected to cover the book's temporal range, and chapters are arranged accordingly in a chronological manner, starting with the early twentieth century and continuing to the present day. The onset of the last century provides not only a convenient starting-point, but it also was a time when certain understandings in the biological and physical sciences had started to provoke profound reflections on the existence of life on other worlds, or the possibility of interplanetary travel. Such considerations found expression in the works of the 'fathers of science fiction', H G Wells and, in France, Jules Verne, beginning in earnest the anticipation of space exploration in Western culture. Although this is primarily a historical book, it seemed futile to foreclose any analysis of contemporary events, particularly as the late twentieth and early twenty-first century witnessed the first Britons to go into outer space, amid other developments. Indeed, as British engagement with outer space is likely to continue in interesting ways, looking towards the future is an important part of this book's final chapter, connecting again with the theme of anticipation that runs through this book. Given this timespan, a range of source materials have been interpreted, some archival in the traditional sense of being held materially in an institutional space, others readily available through online searches. Interviews have been an important source of more personal accounts, including one conducted by the author with Patrick Moore, as part of prior scholarly investigations. A multitude of interviews with key actors from secondary sources has also been invaluable in ascertaining individual outlooks and in triangulating certain events and programmes of work. Published sources, including science-fictional texts, specialist journals, and newspaper reports, have formed a substantial part of the source materials. Furthermore, parts of this book have been abridged from two articles published previously by the author, specifically sections of Chapters 3, 4 and 5.[12]

As a result of this selective sourcing, a number of possible case studies have been overlooked. For example, fictional representations of space exploration, such as C S Lewis' 'cosmic trilogy' of novels, and the *Dan Dare* comic series have been left out, partly because they are the subject of existing works by this and many other researchers, but also because the themes they raised are represented by other examples in the chapters that follow.[13] The primary focus on cultures of space exploration has also tended to result in the omission of some aspects of astronomy from the analysis, such as the advent of radio astronomy, or the ongoing influence of historic observatories, although certain chapters do testify to some of the enduring influence astronomical observation has had in British culture, while there does exist a substantive literature on these topics.[14] In respect to issues of gender, class and ethnicity, the case studies in this book are broadly representative of the cultures and politics of outer space in the UK since the start of the twentieth century. However, this is not to say that they are representative of British society more generally. Indeed, until the penultimate chapter, all the individuals of interest that appear in this book are male, an imbalance that is addressed through discussions of gender in space exploration in Chapter 7. Issues of social class are raised in several chapters, and are seen to have an effect on particular cultural and political engagements with outer space. The question of ethnicity and race presents a stark picture throughout, with the majority of case studies highlighting cultures of outer space in Britain as almost entirely white. This is not to say, however, that issues of race are not discussed at all – parts of Chapter 4 acknowledge the racial tensions that occurred in post-war Britain and their relevance to certain cultures of outer space, while Chapter 5 questions the racial and colonial undertones of outer-space discourse in geopolitical frameworks of space exploration. Such matters of representation raise questions about which parts of society have been able to engage with the cultures and politics of outer space, and while this book does attempt to deal with this question in the ways highlighted above, there remains much work to be done in this area.

Chapter outline

The lives and selected works of two of the most influential writers on outer space cultures in twentieth-century Britain, H G Wells and Olaf Stapledon, are the focus of Chapter 2. Approached through the framework of literary geographies, this chapter considers how the spaces of outer space have been inflected by the authorial experience of place and landscape, and how readers of science-fictional texts incorporated concepts of place into their understandings of outer space. Wells and Stapledon each made important contributions, in the era before the advent of spaceflight, to understanding outer space and humankind's relation to the wider Universe. That these interventions came in fictional form does not diminish their importance – indeed, their narrative format ensured that their concepts of outer space reached as wide an audience as possible in the first decades of the twentieth century, when space exploration was not seen as a viable scientific concept by the vast majority of people. That is not to say that their stories were told in a context devoid of scientific understanding. Both authors were well-connected to some of the leading conceptual developments in the biological and physical sciences of their age, including evolutionism, planetary astronomy and the emerging field of astrophysics. Looking at some of their science-fictional accounts, therefore, highlights the ways in which these emergent fields of knowledge were understood and re-calibrated for audiences in the UK and further afield. The chapter argues that place was instrumental in Wells and Stapledon's formulations of outer-space knowledge, whether referring to the 'rooftop eyries' in which Stapledon became accustomed to viewing the stars of the night sky, to the spaces of science in London's South Kensington where Wells came to learn of the infinite multitude of strange implications for life in the cosmos that were, for him, a logical consequence of evolutionary theory. In looking at the reception of these authors' key science-fictional texts, the chapter further seeks to elucidate the ways in which conceptions of outer space were understood by their audiences, and how readers in different places took different understandings away from these fantastical novels of early-twentieth-century speculative thought.

Chapter 3 turns attention to the paradigm-shifting technology of the rocket, how its potential for spaceflight was recognised by the British Interplanetary Society, and the ways in which it was understood in the emerging context of internationalism during the pre-war period. Having fed on the ideas of science fiction, not only of British writers like Wells and Stapledon, but also from a new wave of writers from America, enthusiasts like Philip Cleator and A M Low became influential in the establishment of the BIS in Liverpool in 1933, and in London by the end of the decade. Here, a combination of unique circumstances, personalities and places combined to cultivate national and international networks of astronautics, the new science of spaceflight that incorporated rocketry, space exploration and human survival in space. As such, the spaces of science in astronautics are examined, including the public houses and domestic residences that were instrumental in shaping this emergent field of knowledge, occurring outside of the official realms of science, and often scorned by established scientific bodies. Furthermore, the ways in which the BIS made connections internationally are explored, thinking through what it means for astronautics to be conceived as an international science, not just in a practical sense, but also conceptually. The chapter concludes by highlighting the culmination of the Society's work in the pre-war period, the 'BIS Space-Ship', a design for a lunar rocket carrying three astronauts intended to land on the

Moon, undertake scientific explorations, and return safely to Earth. While plans such as these were arguably premature, they demonstrate the extent to which members of the BIS were committed to the idea of spaceflight, and the ways in which concepts of interplanetary exploration were connected to the enduring dream of human transcendence.

Chapter 4 draws from recent work in cultural geography, considering the ways in which imaginative representations and lived experience of outer-space phenomena were calibrated in popular culture in the post-war period. Four case studies are investigated: Firstly, the television serial *Quatermass*, specifically the three BBC series of the 1950s, *The Quatermass Experiment*, *Quatermass II* and *Quatermass and the Pit*. Here, while paying attention to the life-world of the screenwriter Nigel Kneale, amid broader social experiences of the Second World War, themes are identified that frame space exploration as an uncertain endeavour, eliciting expressions of fear, awe and the unknown. Similar concepts are approached in the second case study, which is the arrival of the UFO phenomenon in Britain. Here, the first UK government response to UFO sightings is examined, conducted in secret by members of the official 'Flying Saucer Working Party' in 1950, while some of the popular aspects of the phenomenon are also examined, including the formation of the British UFO Research Association. It is suggested that UFO sightings became an integral part of post-war British cultures of outer space, notwithstanding debates about their veracity. One participant in such debates, the astronomer Patrick Moore, forms the nexus of the third case study, which examines Moore's involvement with popular media in the 1950s and 1960s, specifically the BIS publication *Spaceflight* and the BBC television programme *The Sky at Night*. Together, these works helped to promote the science of outer space to a growing audience in the UK. The fourth and final case study in this chapter concerns the early science-fictional writings of Arthur C Clarke, who became one of Britain's most prominent proponents of space exploration, was connected closely to the BIS, and was someone who worked across the two worlds of science fiction and astronautical theory in this period. In examining these case studies, this chapter argues that concepts of space and place were instrumental in understanding and communicating varied cultures of outer space in these contexts, including earthly, aerial, domestic and exploratory spaces.

Chapter 5 broadens the scale of analysis to consider the interactions between geopolitics and space exploration in Britain from the 1950s to the 1970s. This was a time, at the dawn of the Space Age, in which countries around the world were aligning themselves to join the 'space race' between the USA and USSR, including France, Italy, Canada and the UK. The theme of internationalism is returned to here, considering the ways in which the contradictions between nationalism and internationalism were played out with respect to plans for space exploration. The activities of the British Interplanetary Society are examined in this respect, considering the ways in which its members lobbied for a national space programme while taking stock of the UK's shift in geopolitical status from Empire to Commonwealth. A connection to the brief but compelling cultural movement of New Elizabethanism is identified, incorporating the monarchy, aero-mobility and other exciting new technologies of twentieth-century modernity. Ultimately, though, this chapter argues that the UK's material involvement in outer space at this time necessitated a careful calibration between American and European spheres of influence. This negotiation is examined through the case of Ariel 1 – Britain's first satellite – and the European Launcher Development Organisation (ELDO), including the imaginative and material enrolment of the Woomera rocket range in South Australia. In examining multiple modes of representation, including photographic montages depicting international co-operation in space, one of Arthur C Clarke's first novels, and media reports of Britain's early involvement in the exploration of space and the upper atmosphere, the geopolitical aspects of outer space in the UK are examined.

Chapter 6 employs a further shift in scale to examine plans for interstellar exploration that were designed by a team in the British Interplanetary Society during the 1970s, while considering how certain geographical concepts of place and environment can be applied to outer space. 'Project Daedalus' emerged during a period in which the new horizons of space exploration led to more detailed understandings of planets such as Mars and Jupiter, and at a time when social and economic uncertainty was increasingly evident in the UK and the Western world. Ostensibly engaging with 'Fermi's Paradox' of the existence (or otherwise) of alien life in the Universe, Project Daedalus comprised a detailed exposition of the problems, technologies and practicalities of interstellar travel, while also revealing some of the ways in which humankind's place in the Solar System, as well as the wider cosmos, was anticipated. In selecting Barnard's Star, one of the closest stars in the galaxy, as its destination, Project Daedalus also enrolled a broader discourse about scientific authority, the existence of other worlds outside the Solar System, and the constitution of interstellar space. By understanding the Solar System in environmental terms, the project involved detailed considerations of the immediate future of space exploration, with elaborate plans for mining the Jovian atmosphere becoming integral to its implementation. In such ways, Project Daedalus became more than a series of technical papers in the *Interstellar Studies* supplement of the *BIS Journal*, but it helped to frame the scientific community's understandings of outer space, not just of our immediate cosmic environment, but with implications for the ways in which outer space is conceived on a galactic scale.

Chapter 7 examines three episodes in the recent history of British space exploration that have in various ways been connected to questions of nationalism: the story of Helen Sharman, the first Briton in space in 1991; the Beagle 2 Mars lander, whose mission culminated on Christmas Day 2003; and the expedition of Tim Peake to the International Space Station in 2015–16 with the European Space Agency. In recounting the narrative of Helen Sharman's mission to the *Mir* space station, a range of social, cultural and political factors are seen to have come into play. Not only were representations of Sharman inflected by her mould-breaking status as a female astronaut/cosmonaut, but the mission was also characterised by tensions between the worlds of private and public finance in space exploration, the imminent collapse of the Soviet Union, and problematic archetypes of exploration in British culture. By contrast, Beagle 2, a robotic lander attached to the European Space Agency's 'Mars Express' space probe, was promoted with abandon in the British media by its creator Colin Pillinger. While the mission ultimately failed, interpreting the media narratives associated with Beagle 2 reveals much about the national (and nationalist) discourses that became enrolled in outer space culture in this period. The unlikely involvement of the rock band Blur and the celebrated artist Damien Hirst provides a window on the prevailing popular culture of turn-of-the-millennium Britain. The final case study, Tim Peake's mission to the International Space Station, interprets the ways in nationalism was expressed in a variety of performative and symbolic ways, considering events such as the first spacewalk by a British astronaut, and Peake's completion of the London Marathon in space. This chapter brings these three examples together to consider the role of nationalism in contemporary discourses of space exploration in the UK, in terms of its everyday, affective and performative aspects.

The concluding chapter, Chapter 8, reflects on the central research questions of this book, considering the ways in which cultural and political understandings of outer space emerged in Britain since the start of the twentieth century, and how these understandings are relevant in the twenty-first century. Highlighting the main findings of each chapter, a series of reciprocities is identified as key to understanding the geographies of outer space in this context; between science and the imagination, Earth and cosmos, explorative and contemplative modes of engagement. Two final case studies are presented that demonstrate how these themes are extended into the twenty-first century. These are, firstly, the establishment of new space-launch facilities in various parts of the UK, amid the commercialisation and militarisation of outer space, and, second, the conceptual art of Katie Paterson, which draws from the traditions of the sublime in Western culture to consider the nature of outer-space phenomena and the place of humankind in the cosmos. The chapter further reflects on the role of geography in understanding outer space, considering the spectrum of perspectives offered by geographical understandings, while drawing attention to potential new areas of research.

Notes

1 Arthur C Clarke, 'Sir Arthur C Clarke: 90th Birthday Reflections', *YouTube* (9th December 2007) https://youtu.be/3qLdeEjdbWE [15th October 2019].
2 This book primarily uses the term 'outer space' to describe the realm beyond the Earth's atmosphere, conventionally accepted as beginning at the Kármán line of 100km above sea level. Other terms such as 'interplanetary space', 'interstellar space', 'cosmos', and 'the heavens' are used in specific contexts.

3 Denis Cosgrove, *Geography and Vision* (London: I B Tauris, 2008).

4 Steven J Dick, and Roger D Launius, eds., *Societal Impact of Spaceflight* (Washington DC: NASA, 2007); Martin Collins, *Space Race – The US-USSR Competition to Reach the Moon* (Washington DC: Smithsonian, 1999); Martin Parker and David Bell, eds. *Space Travel and Culture: From Apollo to Space Tourism* (Oxford: Blackwell, 2009).

5 Asif Siddiqi, *The Red Rockets' Glare – Spaceflight and the Soviet Imagination, 1857–1957* (Cambridge: Cambridge University Press, 2010); Fraser Mac-Donald, *Escape from Earth – a Secret History of the Space Rocket* (London: Profile, 2019); Frank Winter, *Prelude to the Space Age – the Rocket Societies: 1924–1940* (Washington DC: Smithsonian Institution Press, 1983).

6 Alexander Geppert, ed. *Imagining Outer Space – European Astroculture in the Twentieth Century* (New York: Palgrave Macmillan, 2012); Peter Redfield, *Space in the Tropics – From convicts to rockets in French Guiana* (Berkeley: University of California Press, 2000).

7 Matthew Godwin, *The Skylark Rocket – British Space Science and the European Space Research Organisation 1957–1972* (Paris: Beauchesne, 2007); Peter Morton, *Fire across the Desert – Woomera and the Anglo-Australian Joint Project, 1946–1980* (AGPS: Canberra, 1989).

8 Oliver Dunnett and others, 'Geographies of Outer Space: Progress and New Opportunities', *Progress in Human Geography,* 43 (2019), 314–336.

9 Cosgrove, *Geography and Vision* (London: I B Tauris); Fraser MacDonald, 'Anti-Astropolitik: Outer Space and the Orbit of Geography', *Progress in Human Geography,* 31 (2007), 592–615; Jason Beery, 'Terrestrial Geographies in and of Outer Space', in *The Palgrave Handbook of Society, Culture and Outer Space*, ed. by Peter Dickens and James S Ormrod (Basingstoke: Palgrave Macmillan, 2016), 47–70.

10 Oliver Dunnett, 'Outer Space', in *The SAGE Handbook of Historical Geography, Volume 2,* ed. by Mona Domosh, Michael Heffernan and Charles W J Withers (London: SAGE, 2020), 661–679.

11 Peter Adey, *Aerial Life: Spaces, Mobilities, Affects* (Oxford: Wiley, 2010); Rachel Squire and Klaus Dodds 'Introduction to the Special Issue: Subterranean Geopolitics', *Geopolitics* (2019); Stuart Elden, 'Secure the volume: vertical geopolitics and the depth of power', *Political Geography*, 34 (2013), 35–51.

12 Oliver Dunnett, 'Patrick Moore, Arthur C Clarke and "British Outer Space" in the Mid-Twentieth Century', *cultural geographies,* 19 (2012), 505–522; Oliver Dunnett, 'Geopolitical Cultures of Outer Space: The British Interplanetary Society, 1933–1965', *Geopolitics,* 22 (2017), 452–473 [Permission from SAGE and Taylor & Francis is gratefully acknowledged].

13 Oliver Dunnett, 'C S Lewis and the Moral Threat of Space Exploration, 1938–1964', in *Militarizing Outer Space: Astroculture, Dystopia and the Cold War*, ed. by Alexander Geppert, Tilmann Seiebeneichner and Daniel Brandau (London: Palgrave Macmillan, 2020), 147–170; Oliver Dunnett, 'Framing Landscape: Dan Dare, the Eagle and Post-War Culture in Britain', in *Comic Book Geographies*, ed. by Jason Dittmer (Stuttgart: Franz Steiner, 2014), 27–40.

14 Jon Agar, *Science and Spectacle – the Work of Jodrell Bank in Post-War British Culture* (Amsterdam: Harwood Academic Publishers, 1998); David Aubin, Charlotte Bigg, H Otto Sibum, eds. *The Heavens on Earth – Observatories and Astronomy in Nineteenth-Century Science and Culture* (Durham, NC: Duke University Press, 2010).

2 Science-fictional foundations
A comparative literary geography of H G Wells and Olaf Stapledon

The cultural, political, and fundamentally geographical origins of British understandings of outer space in the modern era can be found in the establishment of a new kind of literature around the turn of the twentieth century, science fiction. While the precursors of science fiction can be traced back much further, its expansion as an increasingly popular form of expression around this time allowed speculative thinkers, including writers and readers, to explore new ideas about outer space, spaceflight, and life in the Universe, that were forged across a variety of actual and imagined spaces. This was instrumental in a period when the science of outer space was not fully established, either in theory or in practice, so speculative accounts took precedence and were important in laying the cultural and political foundations for later developments. This type of fictive imagination emerged throughout the English-speaking and Francophone worlds from the mid-nineteenth century onwards, in the work of writers such as Jules Verne (1828–1905), Edger Rice Burroughs (1875–1950) and Camille Flammarion (1842–1925). This chapter presents a comparative literary geography of two of the most influential British writers in the science fiction of outer space, Herbert George Wells (1866–1946) and William Olaf Stapledon (1886–1950). In doing this, the chapter emphasises the importance of space, place and landscape in the formulation and understanding of concepts of outer space, science and philosophy.

As other scholars have pointed out, Wells and Stapledon were significant because of the positive and widespread reception of their ideas across popular, literary and scientific audiences, and the ways in which their ideas interrelated and complemented each other.[1] Wells is widely known as one of the 'fathers' of science fiction, with his 'scientific romances' such as *The Time Machine* (1895) and *The Island of Doctor Moreau* (1896) becoming international best-sellers throughout the twentieth century, and whose influence in popular culture is still widespread today. Stapledon was lauded among a tighter circle of science-fiction readers and literary critics, writing a more eclectic mixture of spiritual and speculative narratives. While they had contrasting styles and wrote for different audiences, both these authors became formative figures in the minds of British science and technology advocates who were concerned with space exploration from the mid-twentieth century onwards. In a broader sense too, the works of Wells and Stapledon represent a conflation of prominent scientific themes from the first half of the twentieth century, including evolutionism, eugenics and astrophysics. It has been noted that, after the Enlightenment, science fiction emerged as a medium through which existential debates occurred on the place of humankind in the Universe, the role of the supernatural, and the relationship between the human and non-human.[2] This chapter will explore this contention in demonstrating that British outer-space literature provided an ideological testing-ground for ideas of modernity in the twentieth century, while being instrumental in shaping the subsequent cultures and politics of outer space in the UK. From a distinctly geographical perspective, it examines some of the perceived implications of evolutionary theory for the future of humankind, the relevance of new understandings of the mutability of the cosmos, and the ways in which expanded concepts of spirituality and the sublime challenged the very meaning of the Universe at the beginning of the twentieth century.

Literary geographies and science fiction

Foregrounding the importance of space, place and landscape in understandings of outer space requires an appreciation of the well-established conceptual links between geography and literature that extend from the mid-twentieth century to the present day. Whereas early literary geography has been criticised for blunt or overly descriptive analyses of place in literature, recent engagements have seen the emergence of more analytical, reflexive accounts of the relationships between space and literary themes, drawing from the 'cultural turn' in geography, and connections with critical theory including psychoanalytic and materialist concerns.[3] Such accounts, and their precedents in areas such as postcolonial criticism, refer to the full spectrum of relations between author, reader and text, considering the spaces of writing and influence in the creation of literary texts, as well as the reception of texts across various reading positions and divergent audiences.[4] Spaces such as the poet Dylan Thomas' writing shed on the Taf Estuary in South Wales, and the walking spaces of the English novelist Arnold Bennett, have been accorded particular significance in processes of writing, while spaces of reception, presentation and collaboration also feed into the intricacies of the new literary geography.[5] There has also been a renewed appreciation of the conceptual spaces created within literary texts, considering the ways in which these imaginative geographies are co-constituted by authors and readers, and the impact they have in various understandings of the world. Thinking methodologically, researchers have pressed for a blending of approaches from literary criticism and critical geography, with the precise analytical frameworks of the former complementing the expansive relational analyses of the latter.[6]

Literary geographies of science fiction have received some limited scholarly attention, most significantly through the work of Rob Kitchin and James Kneale, who have considered the ways in which space has been configured in the science-fictional sub-genre of 'cyberpunk' in the late twentieth century.[7] Kitchin and Kneale argue that, since science fiction is by definition a non-realist form of literature, it represents 'a privileged site for critical thought', elements of which are 'highly geographical' in nature.[8] In this way, speculative science-fictional spaces can help in the conceptualisation and understanding of possible other worlds or altered future states in a very real sense. Kitchin and Kneale draw on science-fiction scholar Darko Suvin's concept of 'cognitive estrangement', whereby narrative engagement with a form of radical alterity, or technological 'novum', such as time travel or instantaneous communication across space, is used as a critical tool for imagining alternate realities.[9] In this way, rather than being dismissed as a niche interest, science fiction opens up the potential for critical commentary on contemporary societies, perhaps to be considered as 'an act of defiance, a literature of subversion, not whimsy'.[10] Hence, science fiction can be considered as 'more than a simple *negation* of realism and rationality', rather its narrative ambiguities help make sense of lived experience in broadly relatable ways.[11] By extending this approach to cognitive estrangement, and routing it through the work of new literary geographies, the spaces of science-fiction literature can be considered in a reciprocal sense, not only through the potential of imaginative spaces to alter conceptions of lived reality, but also through the ways in which certain spaces, places and landscapes become instrumental in imagining the conditions of radical alterity. It is through such avenues that geographies of science fiction can be explored with useful critical intent.

While existing work in this area has productively interpreted geographies of postmodernist science fiction, such methodologies have been applied less commonly to earlier 'modernist' science fiction, the central tenets of which have been the idea of spaceflight and the imaginative possibilities of other worlds. Tracing the origins of these speculative concepts, historical studies have demonstrated how some of the early progenitors of science fiction, such as Johannes Kepler's *Somnium* (1608), were associated with revolutionary Enlightenment-era understandings of the Universe, ideas that became central to a broad range of modern philosophies. In Kepler's work, for example, imaginative descriptions helped to re-conceptualise bodies such as the Moon away from Classical understandings of immutable 'heavenly spheres' to being seen as 'other worlds' with many similarities to our own Earth.[12] This was a cognitive estrangement of the most profound kind. In the nineteenth and early twentieth centuries, the establishment of a distinctly British genre of scientific romance explored the imaginative potential of new ideas in evolutionary theory and astrophysics. This new category of literature made its mark as 'the romance of the disenchanted universe', often cutting against the social and political grain of British Victorian society.[13]

In this way, the themes of cosmic ascent (an imaginative forerunner of human spaceflight) and the speculative spaces of other worlds, which are central to understanding geographies of early-twentieth-century science fiction, open up a window on radical alternative configurations of society at a particular place and time.

Understanding outer space in early-twentieth-century Britain as an emergent space, onto which experimental ideas could be projected, adds a further dimension to studies in literary geography, encouraged by pioneering work in the geographies of science fiction. As such, this chapter investigates the outer-space fiction of H G Wells and Olaf Stapledon in relation to certain biographical episodes, places of literary significance, and their initial critical reception. Although they formed part of a wider coterie of scientific romance authors in this period, including names such as George Griffith (1857–1906), John Davys Beresford (1873–1947), and Sydney Fowler Wright (1874–1965), Wells and Stapledon stood out from this group in terms of their broad appeal and influence. Furthermore, it has been observed that '[Stapledon's] fiction is fundamentally different from Wells's, however much its ideas may have been fostered by Wells's historical and utopian speculations'.[14] These differences are understood most clearly in terms of narrative form: whereas Wells was a master of storytelling, his works following established structures of narrative writing, Stapledon tended to forego novelistic conventions, preferring instead a more descriptive form of writing that extended conventional representations of time and space. Indeed, Stapledon's writing was characterised by the most expansive of themes, such as the evolution of humankind into the far future in *Last and First Men* (1930), and the myriad possibilities of alternative forms of consciousness across the Universe in *Star Maker* (1937). These texts, his most lauded works, arguably had more in common with utopian treatises, fantasy travelogues or historical expositions than standard types of fictional writing, with Stapledon himself reluctant to characterise them as novels at all. Wells had a much more voluminous output, writing science-fictional and conventional novels on a variety of themes. Those of his works that incorporate interplanetary themes most clearly, however, are *The War of the Worlds* (1898), the archetypal Martian invasion story, and *The First Men in the Moon* (1901), recounting a comic-imperialist journey to the Moon. The remainder of this chapter turns to the lives, spaces and audiences that together help to formulate a comparative literary geography of these two writers, with respect to their most prominent examples of outer-space fiction.

Life-paths: biographical cultures of science

Comparing the biographies of H G Wells and Olaf Stapledon can help to elucidate the ways in which their respective life-paths instilled in each writer the capacity to make sense of the Universe in particular ways. This approach draws on scholarship that has recognised 'the intersection of the geographical and biographical' in researching aspects of self, place and identity in a variety of historical contexts.[15] Similarly, others have commented on how, in order to make sense of the different spaces people occupy throughout their professional and personal lives, an approach focussed on 'geographical biography' would yield insightful accounts of scientific knowledge creation, an approach that could just as effectively be applied to understanding literary narratives.[16]

Accordingly, while looking to make sense of Wells and Stapledon's life-long fascination with extra-terrestrial space, biographers of both writers have attributed significance to childhood experiences of astronomical observation, episodes that emphasise the importance of place and landscape. The young Wells, while spending his holidays at Uppark estate in West Sussex, where his mother worked as a housekeeper, discovered and assembled a telescope in an attic room that had belonged to the estate's former owner Sir Harry Fetherstonhaugh, staying up into the early hours 'inspecting the craters of the moon'.[17] When looking through the telescope across the landscape by day, he noticed that 'it showed everything upside down' – a common quirk of astronomical observation when using certain types of telescope – but when turned to the sky at night, the inversion made no appreciable difference due to the lack of a conventional sense of 'up and down' in space.[18] This acts as an effective metaphor for Wells' outlook on the speculative potential of outer space. Furthermore, this vision is said to have 'lifted [Wells] out of himself and placed him within a cosmic order', allowing an imaginative escape from the social, cultural and political restrictions of late Victorian society.[19] Similarly, Stapledon is accorded his own formative moment at the end of a telescope as a child in Port Said, Egypt, where his father was working for an international shipping company. From a balcony at home, 'the clear, wide-horizoned desert sky' gave way to a sense of 'star-struck wonder' at the immensity of the cosmic realm.[20] In 'one divine hour never forgotten', Stapledon is said to have observed the contours and mountains of the Moon, considering for the first time the presence of other worlds.[21] In these accounts, the twinned technological cultures of amateur aristocratic astronomy and maritime telescopic observation helped to evoke a common sense of the astronomical sublime, incorporating themes of cosmic ascent and the contemplation of other worlds. Together they point towards not only the creative significance of place – the attic room and the desert balcony – but also the wider cultural and political circumstances of place, in terms of social class, empire and technology.

Both writers benefited from extended educational opportunities that became pivotal to their literary perspectives on outer space. Following some initial experience as a school science instructor, Wells secured an undergraduate scholarship for a degree in biology at the Normal School of Science (later Imperial College London), graduating in 1890. Here he studied for one year under 'Darwin's Bulldog' T H Huxley, where scientific cultures of evolutionism, including emerging theories of eugenics, left a lasting impression.

As a student, Wells saw the potential of science to change society fundamentally, even on a global scale, a realisation that formed a welcome alternative to the fixed view of the world that was engrained in the conservative
religiosity and traditions of his upbringing.[22] Stapledon, having graduated
from Balliol College, Oxford, with a degree in modern history, completed a
PhD in philosophy and ethics at Liverpool University in 1929. Here, 'he drew
from a rich brew of ideas and provocative questions', reading philosophers
of science and epistemologists, such as Henri Bergson, William James and
William Winwood Reade.[23] At Liverpool, Stapledon embedded himself in
the discourse of modern science, particularly the dynamic fields of evolutionary and eugenical theory, where Darwin's insights were spreading into areas
of social and political significance. In subscribing to the journal *Nature*,
Stapledon also kept himself informed on developments in other sciences
such as astrophysics, where researchers such as James Jeanes and Arthur
Eddington were beginning to understand aspects of the true complexity of
the Universe. Steeped in this intellectual environment, Stapledon developed
a philosophical stance that has been seen as central to understanding his science fiction. In this, he maintained that older convictions of morality in society had been 'eroded' by modern developments in cosmology, psychology and
the life sciences, and needed to be replaced by a 'modern theory of ethics'.[24]
For both writers, combining educational influences in the fields of culture,
morality and science with the imaginative possibilities of extra-terrestrial
spaces would help to engender rich literary geographies of other worlds.

An emerging sense of the potential of science to change society not only in
Britain, but also around the world, led both Wells and Stapledon to become
committed socialists and advocates of a world-state. Wells was 'acutely aware
of social class in British society', his upbringing exposing him to the precarity
of life in late Victorian England for those not born into wealth.[25] As a student, he attended Fabian Society meetings at the home of William Morris
in Hammersmith, London, uncommonly perceiving a connection between
science and socialism, in working towards what he called 'a planned inter-co-
ordinated society'.[26] At the same time, it has been noted that the British
Empire 'would shape Wells's vision of a unified world', his adult life centred
on the imperial metropole of London.[27] Comparably, Stapledon's early life
occurred in the shadow of the British Empire at its broadest extent, as he witnessed the coming and going of all manner of people and goods in and out of
Port Said, at the northern terminal of the Suez Canal, and Liverpool, where he
had returned with his parents in 1901. Although more secure in his middle-
class mercantile roots, Stapledon immersed himself in the working-class cultures of north-west England, exemplified by his long-standing voluntary role
teaching history, philosophy and politics for the Liverpool branch of the
Workers' Educational Association. Stapledon's role in the First World War is
also significant, and as a life-long pacifist, he served as a driver for the Friends'
Ambulance Unit, a Quaker organisation operating near the front lines in
France and Belgium. Hence, the two writers became committed to socialism

through different routes: Wells by a sense of unfairness in witnessing the bourgeois elements of Victorian society in comparison to his own upbringing, and Stapledon in recognising the plight of the working classes from a more secure social perspective, provoking a sense of guilt that stayed with him throughout his life that fed his desire to make change.

With Wells' career as a writer taking off roughly three decades ahead of Stapledon's, his work became part of the wider cultures of popular fiction from which Stapledon naturally drew considerable influence. In 1931, Stapledon initiated a correspondence with Wells that was to last over a decade, following the release of his first major work of fiction *Last and First Men* (1930). In this, he commented on reviewers' comparisons with Wells, acknowledging his influence with striking candour:

> Your later works I greatly admire [...] Then why, I wonder, did I not acknowledge my huge debt? Probably because it was so huge and obvious that I was not properly aware of it. A man does not record his debt to the air he breathes in common with everyone else.[28]

This correspondence led to a meeting in 1936, for lunch at Wells' house in London, during which the two writers aimed to 'settle the Whole Damn Silly Universe'.[29] Although clearly, they shared much common ground, Stapledon afterwards began to question Wells' ideas, 'both privately and in print', criticising his conceptions of economics and spirituality and their role in a socialist world system.[30] Both commented on the state of world affairs with opposing degrees of despair and hope, with Wells questioning the very role of the writer during the tribulations of the Second World War and Stapledon frequently speaking out in local and international forums, in later years being described as 'the Cold War's peace pilgrim'.[31] One letter in their correspondence stands out, in which Stapledon counters Wells' claim that he is caught up in proletarian and spiritual matters, in the form of a cartoon. Wells is caricatured causally walking his own path between these two caged vices, while a bird watches on. Stapledon comments, 'I am not really in either of the cages, believe me! I am the jackdaw, free but uncertain.'[32] The relationship between the two writers has been summed up by science-fiction critic Robert Crossley, who characterises Stapledon as an errant disciple of Wells, growing apart from his mentor towards the end of their lives. These contrasts between the two writers' life-paths, as well as aspects of their lives that they had in common, particularly their political and social perspectives, provide essential context for understanding their imaginative projections of outer space.

H G Wells: science, society, and interplanetary space

Wells' scientific romances, concentrated within the early part of his writing career, experimented with ideas of time, space and what it means to be human. An integral part of these stories was the way in which the speculative elements were extrapolated from a recognisable sense of place and culture. In doing this, Wells connected his experience of everyday life in lower-middle-class England with the emerging popular science of the era, in the context of his left-of-centre political views, striking a formula that was to stand the test of time in science-fiction literature. As Robert Shelton has put it, Wells 'united, in a binocular vision, the scientific (biological) and the sociological (utopian) perspectives of the Victorian and post-Victorian eras', offering a unique insight into the cultures and politics of the time, and helping to demonstrate how certain characteristics of place informed Wells' literary geographies.[33]

This connection between speculative narratives, observed science, and the spaces of everyday life in Wells' work is perhaps exemplified most clearly in *The War of the Worlds*. This novel, which is told from the perspective of an ordinary man witnessing a belligerent Martian invasion, is set in the regional locales of south-east England and London, from the Thames Valley towns of Richmond and Putney to the further suburbs of Woking and Addlestone. Some of these sites were personally scouted by Wells on a bicycle, and he later recalled how he 'wheeled about the district [of Woking] marking down suitable places and people for destruction by my Martians'.[34] As well as clearly being situated in this particular part of southeast England, the novel's relatability came through its integration with the scientific discourses of the time, including the 'inhabited Mars' thesis of the late nineteenth century and ongoing debates around evolutionary theory. In this respect, Wells opened his novel with a discussion on human complacency and the possibilities of extraterrestrial life:

> Men like Schiaparelli watched the red planet [...] but failed to interpret the fluctuating appearances of the markings they mapped so well. All that time the Martians must have been getting ready.[35]

As this passage indicates, astronomical science in the late nineteenth century became wrapped up in public debates about the possibility of life on Mars. A central element of this discourse was the emergence of new 'Mars maps' by astronomers Giovanni Schiaparelli in the 1870s and Percival Lowell in the 1890s, representations that appeared to make the case for large-scale irrigation works on Mars.[36] Rather than acting as scientifically 'truthful' documents, maps such as these were both inflected with, and received in the context of, broader cultural and political discourses, into which Wells inserted his novel. Also of relevance in the broader public discourse around science at this time was the consolidation of evolutionary theory following Charles Darwin's death in 1882, and its infiltration into social, cultural and political arenas.[37] For example, researchers have examined extensively the hermeneutics of evolutionary theory in late-nineteenth-century literature, demonstrating the permeability between scientific and popular cultures during this period, and explaining how popular culture became a medium through which scientific theories were understood by broader public audiences.[38] As part of such work, certain critics have outlined the ways in which evolutionary theory was written into *The War of the Worlds*, considering it as a reflection of social Darwinism, as a future projection of the role of technology in evolutionary development, and on the place of eugenics in society.[39] In such ways, Wells' work was intimately associated with the prevailing cultures of science in turn-of-the-century Britain.

In this respect, Wells' tutelage under Thomas Henry Huxley, and his school of 'South Kensington Biology' appear to have been central. Huxley was a leading member of 'The X Club', a group of scholars who sought to professionalise biology following the Darwinian revolution of the mid-nineteenth century, meeting on a monthly basis in London from 1864 to 1893.[40] As part of this professionalisation process, Huxley recruited students who were training to become school science teachers, finding in Wells such a willing and able disciple. Central to Huxley's method was the laboratory, located in the newly designed 'science schools' building in South Kensington, along with apparatus, such as the microscope, the purpose being to propagate a particular kind of 'simple, synthetic and assimilable' science.[41] Wells later described this learning environment:

> Here were microscopes, dissections, models, diagrams close to the objects they elucidated, specimens, museums, ready answers to questions, explanations, discussions.[42]

Researchers have further suggested that the planned layout of Huxley's laboratory space, which benefited from skylight illumination on the top floor of the building, was instrumental in fostering the modern learning environment required to teach the new Darwinised biology of the late nineteenth century.[43] Of broader relevance was the surrounding scientific environment of South Kensington, with educational institutions and museums developing in tandem as part of the professionalisation of science in this period. In learning from Huxley and his demonstrators in the laboratory, the lecture theatre, and the wider institutional environment, Wells came to realise that '[t]he mechanism of evolution [was] a field for almost irresponsible speculation', and indeed in *The War of the Worlds*, Wells was among the first to speculate that 'the Darwinian struggle for existence [was] something which could extend across interplanetary space'.[44]

In *The War of the Worlds*, Wells connected evolutionary theory to themes of imperialism, as part of this broader speculative endeavour. Having stated in Darwinian terms in the opening lines of the novel that 'life is an incessant struggle for existence', Wells goes on to suggest that the coming Martian invasion might represent some sense of natural order in the Solar System.[45] This reflects some of the societal tensions that evolutionary theory exposed, entanglements that 'confuse[d] the standards and definitions that configure modern society and self-identity'.[46] Here, thinking back to the young Wells looking through the telescope at Uppark, conventional standards become upended when looking speculatively towards the other worlds of the Solar System. Thus, Wells inverts the hubris and belligerence that he associated with humankind, asking:

> Are we such apostles of mercy as to complain if the Martians warred in the same spirit?[47]

Here, and throughout the novel, humans are definitively downgraded, having been corralled and enslaved by the Martian invaders. When, at the end of the novel, the Martians themselves are defeated, it is through the action of microscopic bacteria, rather than at the hands of any conventional hero of humanity. In such ways, the narrative of *The War of the Worlds* is situated distinctly in the climate of scientific and political debates of the late nineteenth century, with Wells channelling Huxley's strident arguments on evolutionism in the context of the 'inhabited Mars' discourse, while also outlining a nascent anti-imperialism that was to become a feature of his later works. Over and above acting as narrative backdrops that would be familiar to some readers, the geographies of scientific London and the suburban features of the English home counties are seen to have actively contributed to the overriding themes of the novel. At the same time, a wider geographical context including the circulation of Mars maps, the seeming omnipotence of the British Empire, and the spatial aspects of evolutionary theory all fed into Wells' science-fictional vision.

Similar topics are explored in *The First Men in the Moon*. In this, one of Wells' later scientific romances, two characters set off on a journey to the Moon, their spherical spaceship powered by a gravity-repelling substance. While themes of colonial plunder are central to this novel, it is perhaps Wells' depiction of the Moon-dwelling, insect-like 'Selenite' species that is most interesting, against which interpretations of biological science and social politics can be read. Towards the end of the novel, Wells devotes a chapter to 'The Natural History of the Selenites', in which a complex society is described. Sherborne interprets this imagined society has been interpreted as 'superior to ours in some respects, yet a ridiculous exaggeration of ours in others'.[48] In Selenite society:

> every citizen knows his place. He is born to that place, and the elaborate discipline of training and education and surgery he undergoes fits him at last so completely to it that he has neither ideas nor organs for any purpose beyond it.[49]

Here, Wells deviates from theories of eugenics that were gaining popularity in the early twentieth century, whereby certain less-desirable facets of society could be eliminated benignly through selective breeding. Such accounts, promoted by eugenicists such as Francis Galton, were seen by

Wells to reduce certain kinds of people to the status of cattle, reproducing the class-based rationales for societal organisation that he was strongly against.[50] Instead, Wells' Selenites in *The First Men in the Moon* constitute an inter-dependent, individually specialised society, where social role and biological form are integrated closely, so that 'with all sorts and conditions of Selenites – each is a perfect unit in a world machine....'.[51] Wells' use of an ellipsis at the end of this statement is perhaps an invitation to the reader to consider the actual possibilities of such a social system, a suggestion that tallies with Wells' active engagement with utopian thinking and socialist world politics throughout his life. In any case, through looking at aspects of both *The War of the Worlds* and *The First Men in the Moon*, we can see that, for Wells, writing about interplanetary space speculatively, including the sublime possibilities of life across the Solar System, implicated various hopes and fears about contemporary society that were engendered through the specific political, cultural and spatial contexts of turn-of-the-century London.

Olaf Stapledon: landscape, cosmos, and the sublime

While the broad importance of early-twentieth-century science and society can also be recognised in the works of Olaf Stapledon, an appreciation of specific, place-based inspiration is particularly significant in understanding this writer's imaginative narrations of the cosmos. Stapledon's most celebrated works of fiction, *Last and First Men* and *Star Maker*, were written in the late 1920s and early 1930s, when he was living with his wife Agnes and two children, Mary and John, in a house in the coastal town of West Kirby in north-west England. It has been noted that, for Stapledon, 'the provincial and the cosmic were no antimony', indicating not only that he drew creative solace from his surroundings, but also that he perceived a sense of the sublime in the regional landscapes that helped fuel his literary imagination.[52] When considering Stapledonian themes such as contemplation of humanity's long-distant futures, the physical limits of the Universe, or the nature of consciousness, one has to question how such a provincial sense of place and landscape informed his ideas of cosmic sublimity. Some of the answers can be found in Stapledon's experiences of place and landscape in the region where England meets Wales and the Irish Sea.

Before he came to live in West Kirby, Stapledon was a frequent visitor to his parents' house, 'Annery', which his father had built in the same town in 1910. Here, in a 'rooftop eyrie', a telescope was installed, where he was said to have spent many a night observing the stars, echoing his earlier childhood experiences in Port Said.[53] The significance of such activities becomes especially notable when considering a modification made by Stapledon some years later on his own house at 7 Grosvenor Avenue, West Kirby. This installation consisted of a large opening in the ceiling of the attic space that was his writing room, across which would roll a retractable panel, forming an open skylight that afforded direct views of the heavens above. The connection that Stapledon craved with the cosmic realm thereby found material form in his personal writing space, in a manner not unlike some contemporary 'skyspace' art installations that provoke affective encounters with open vertical spaces.[54] Crossley also notes the importance of Stapledon's sky-hatch, both creatively and spiritually, suggesting that:

> His imagination thrived in sunlight and starlight, and to get a taste of the freedom of infinite space, he permitted himself a single luxury [...] in the chapel of his attic-observatory.[55]

Clearly then, for Stapledon, the creative writing process was not only 'a personal pursuit conducted in a private space', but it also required the material transformation of a usually mundane and unused domestic space, both in terms of its conversion into a writing room, and through its direct connection to the open skies.[56] Indeed, more than just enabling direct observation of the skies above, this arrangement also seemed to foster an almost spiritual or transcendent connection to the cosmic realm.

Whereas the tranquillity of suburbia in the 'comfortable obscurity of the North' was an important facet of Stapledon's literary geography, wider interactions with landscape and place were also influential.[57] Personal diaries and later reminiscences reveal that he took holidays with his family along the Welsh coast in the years leading up to the publication of *Last and First Men*. It was while walking along a cliff-top coastline, noting a cemetery near the edge of the precipice, and observing a group of seals lounging on rocks in the sea below, that Stapledon is said to have been struck by his vision of humankind evolving over the course of many ages. Writing in 1954, he recalled:

> Long ago (it was while I was scrambling on a rugged coast, where great waves broke in blossom on the rocks) I had a sudden fantasy of man's whole future, aeon upon aeon of strange vicissitudes and gallant endeavours in world after world [...] since then, year after year, I have tried to create in words symbols of that vision.[58]

Later analyses have re-constructed the details of this visionary experience, attributing it to a combined memory of seal-watching near St. David's Head in Pembrokeshire in 1926, and cliff-walking on the isle of Anglesey two years later, during which Stapledon is said to have contemplated 'the splendour, the oddity, and the fragility of all mind and life'.[59] Here, a combined impression of cliff-face rock strata, buried human remains, the 'humanity' of the seals, and the destructive power of the sea induced a sense of sublime wonder in Stapledon. As well as being associated conventionally with understandings of 'the heavens' and spiritual ascent, notions of the sublime can, in this context, be connected to understandings of geological time, which, since the nineteenth century, had transformed narratives of the history of the Earth and of humankind, both in scientific and fictional representations.[60] Themes of deep time and the nature of humanity are central to *Last and First Men*, which narrates the future development of eighteen distinct human species over the course of several millennia, including airborne varieties and 'seal-like sub-men', across the diverse planetary spaces of the Solar System.[61] In this case, specific experiences of place informed an important part of Stapledon's *oeuvre*, in a time in which frameworks of understanding the Earth's place in the cosmos were being extended in fundamental ways in both the geological and biological sciences.

The opening pages of *Star Maker* draw similarly from an intriguing mixture of scientific, literary and geographical influences. While often lauded for its originality, the text is a modern echo of the age-old literary genre of the visionary cosmic voyage, in the manner of Cicero's *Dream of Scipio* from the first century BC.[62] Stapledon's expansive narrative begins in a modest setting whose imaginative origins can be traced to the landscapes of the Wirral peninsula, where Stapledon lived in West Kirby. This is a small town overlooking the expanse of the Dee estuary, and onwards to the Welsh coast and Cambrian Mountains. Stapledon is known to have frequently climbed Caldy Hill, a modest ascent of some seventy metres, from paths out of West Kirby, to observe this panorama. The nearby Stapledon Wood has since been named for his long association with this area. From such a place, *Star Maker* begins:

> One night when I had tasted bitterness, I went out on to the hill. Dark heather checked my feet. Below marched the sub-urban street lamps. Windows, their curtains drawn, were shut eyes, inwardly watching the lives of dreams. Beyond the sea's level darkness a lighthouse pulsed. Overhead, obscurity.[63]

From here, a star, and then the entire night sky, are revealed to the narrator, who begins a visionary, dis-embodied journey through the centre of the Earth, out the other side and onwards into the cosmic realm, observing the Earth receding away to become just another speck of light in the darkness. A sense of existential anxiety is described on the hill's summit:

> One tremulous arrow of light, projected how many thousands of years ago, now stung my nerves with vision, and my heart with fear. For in such a universe as this what significance could there be in our fortuitous, our frail, our evanescent community?[64]

Here, a clear sense of the poetic sublime is evoked, in both the contemplation of the origins of distant starlight, and also through the affective resonance of experiencing this simple astronomical phenomenon, articulated through a feeling of fear. Having inexplicably been transported into outer

space, the narrator describes further astrophysical phenomena that codify
Star Maker as a specifically modern cosmic voyage:

> Very soon the heavens presented an extraordinary appearance, for all
> the stars directly behind me were now deep red, while those ahead of me
> were violet. Rubies lay behind me, amethysts ahead of me.[65]

Here, the phenomena of spectroscopic redshift and blueshift are described,
which were recognised in early-twentieth-century astrophysical science
to account for the movement of light sources away from and towards the
beholder at great speeds. Not only does this situate *Star Maker* in the active
scientific discourses of the time, but it also tells us about Stapledon's conception of the Universe more generally. The explanation of redshift was an
integral part of Hubble's Law in 1929, which provided the theoretical basis
for an expanding Universe. This occurred in the context of Albert Einstein's
general theory of relativity, which helped to replace prior notions of a fixed
or static Universe with a more dynamic cosmos in which the relationship
between time and space was seen as mutable, open in Stapledon's mind to
the myriad possibilities of extra-terrestrial life, the wonders of other worlds,
and even the notion of sentience in stars and planets.

Stapledon's ideas about interplanetary evolution and conceptions of a
dynamic Universe are further explored in *Star Maker*, with the theme of symbiosis central to much of the narrative. In contrast to *Last and First Men*,
the condition of humanity refers not only to the people of Earth and their
descendants, but also to various forms of life in the Universe that have attained
a certain level or rank of consciousness, incorporating myriad differences in
physical form, mental status, and spiritual attainment. These variances partly
relate to the physical characteristics of the planets themselves, spread across a
multitude of galaxies, with environmental factors such as atmospheric density
and gravitational mass playing formative roles in the imagined evolutionary
process. Most significant to the narrative of *Star Maker* are human-class species that evolve towards symbiosis, as seen on one planet where sea-dwelling
'ichthyoids' live in close co-operation with crustacean-like 'arachnoids', having 'moulded one another to form a well-integrated union', including a type
of 'biochemical interdependence'.[66] More than just living symbiotically, they
achieve a form of social utopia, as described by the narrator:

> We learned also with wonder that at the height of the pre-mechanical civili-
> zation of this planet, when in our worlds the cleavage into masters and eco-
> nomic slaves would already have become serious, the communal spirit of
> the city triumphed over all individualistic enterprise. Very soon this world
> became a tissue of interdependent but independent municipal communes.[67]

Here, aspects of evolutionism and socialism are imaginatively intertwined
in a state of 'personality-in-community' which Stapledon considered to be

a fundamental goal of human society.[68] The state of symbiosis that is developed in *Star Maker* thereby finds partial expression through Stapledon's vision of societal harmony, on the one hand, and a form of domestic harmony on the other, as the narrator returns home to his 'little glowing atom of community', reconciled, at the end of the narrative.[69] For Stapledon, therefore, the personal and the cosmic were connected deeply.

Throughout *Last and First Men* and *Star Maker*, it is apparent how a series of spaces, from Stapledon's home and writing room, to the surrounding landscapes of north-west England and Wales, formed an integral part of his literary geography, in the context of the broader cultures of science from which Stapledon also drew inspiration. Also significant has been an appreciation of various forms of the sublime, inflected by the contemporary scientific discourses of astrophysics and geology. Through looking at aspects of Stapledon's literary geographies alongside those of Wells, we can see how place matters in British cultures and politics of outer space in the early twentieth century, both in terms of its specificities and its broader contexts.

Readings of the Universe, Earth, and humanity

In a commentary on Olaf Stapledon's *Star Maker*, the writer and critic Bertrand Russell stated that, '[w]hen confronted with the vastness of the cosmos and the smallness of our planet, different men react very differently'.[70] Such differences relate to not only the literary imaginations of writers like Stapledon and Wells, but also to their readership, some access to which can be gained through investigating published reviews of their works. Although not by any means fully representative of the ways in which these books have been read and understood, through their analysis, some sense of the formulations of knowledge and attitudes about outer space in early-twentieth-century Britain can be gained, including the role of space and place in such formulations.

Wells and Stapledon both have been acclaimed as science-fiction writers, in the UK and around the world, but the depth and range of their appeal have varied considerably, both during and after their lifetimes. These differing qualities in reception can be traced to a variety of factors, including the ways in which certain narrative themes have played out over time, as well as differences in cultures of reading, and styles of writing. Wells' fiction initially tended to consist of short, serialised stories for monthly periodicals such as *Pearson's Magazine* and *The New Review*, which would garner an established audience among casual readers, before being updated and released in full-length novels. He was able to tap in to the growth of middle-ground fiction in late-Victorian Britain, occupying the intellectual space between the cheap and popular 'penny-dreadfuls', and the more high-brow and expensive three-decker-novels.[71] This appealed to an emerging mass-market at a time of near-universal literacy, with reading having been encouraged on a national scale following the UK Public Libraries Act of 1850.

As Wells put it himself, '[t]he habit of reading was spreading to new classes with distinctive needs and curiosities'.[72] His productivity matched this eager market for new and original stories, as he completed twelve novels between 1895 and 1905, establishing a reputation as one of the best-known writers of the new scientific romances. Stapledon, by contrast, developed his writing style gradually, his expansive, some might say ponderous, style not lending itself well to serialisation. He published relatively few books – ten over the course of his life – starting at a time when the global economy was slowing, and conditions for writers were much less favourable. The advent of the Second World War further stymied Stapledon's productivity, with paper shortages and other restrictions having an effect.[73] In the post-war period, some have attributed a decline in Stapledon's stock to the ethical tone of his works, with themes such as eugenics quickly turning out of favour in public discourse.[74] As such, whereas *The War of the Worlds* garnered an initial print run of 10,000 in 1898, with multiple editions soon to follow, *Star Maker*'s first edition ran to just 2,500 in 1937.

The War of the Worlds has a famous and well-documented history of re-invention right up to the present day, having been adapted into countless films, television and radio serials. These re-interpretations began almost immediately, with unauthorised serialisations in the New York *Evening Journal* and Boston *Post* attracting Wells' ire for changes made 'in order to fit it to the requirements of the local geography' of the north-eastern United States.[75] A 1906 French translation contained a series of distinctive pencil-and-ink illustrations of the Martian machines by the Brazilian artist Henrique Alvim Corrêa (1876–1910), extending the novel's international appeal (Fig. 2.1). Most famously, *The War of the Worlds* was broadcast in a 1938 radio production by Orson Welles, reportedly resulting in mass panic throughout the United States at the prospect of an actual and imminent Martian invasion.

Such interventions speak to the novel's enduring impact on the public imagination, aspects of which can be traced to its initial reception in the UK press. Indeed, the first reviews of *The War of the Worlds* generally contained high praise for Wells' second scientific romance, notably stimulating discussion on a range of social and scientific issues. *The Daily News*, a liberal London paper, praised the book for its modern scientific outlook, stating that Wells 'possesses not only an imagination that is exceedingly powerful, but also the mind that is essentially scientific'.[76] The *Pall Mall Gazette* presented Wells as 'our authority' in natural science and astronomy, and such was the extent to which *The War of the Worlds* brought speculative science to life, that 'we readily and gladly accept [his fiction] as fact', the review in question partly written as faux-reportage to the events depicted in the novel:

> The first shot struck earth on Horsell Common, and others kept arriving between there and London until there were ten here altogether, the last one striking and crushing deep into the ground a large villa at Sheen.[77]

Figure 2.1 Illustration by Henrique Alvim Corrêa, from the 1906 Belgium (French language) edition of The War of the Worlds. The original caption stated: 'Martian Fighting Machine in the Thames Valley'

Source Credit: Henrique Alvim Corrêa/Public domain

Similarly, the *Globe* reported that 'Mr. Wells has made his mark [and] is happiest, unquestionably, when his scientific acquirements come to the aid of his natural powers of invention'.[78] In a reflective review, *The Graphic* also focussed on scientific themes, suggesting that the portrayal of the Martian invaders 'upsett[ed] all conventional imagination'. The reviewer commented on the 'trains of thought' that Wells had instigated on the nature of life and evolution:

> [H]e is ready to allow natural selection to have produced as its highest product some wonderful insect, or even some amorphous creature, all brain and no body, for mechanical appliances we may believe are to supersede limbs in time even on this world.[79]

Here, a sense in which natural selection might entail fundamental changes to the human body of the future is provoked, indicating that speculative reflections about evolution and the nature of humankind were a substantial part of the impact *The War of the Worlds* had on its immediate readership. Another reviewer in the *Morning Post*, noting how Wells 'shakes our credulity severely' in the novel, went on to discuss from a matching critical stance the nature of the Martian invaders:

> [E]verything is given up to brain power [...] In place of the insignificant arms and legs, they possess tentacles which combine the flexibility of a monkey's tail with the sensitiveness of a beetle's antennae.[80]

In this case, while criticising the outlandishness of Wells' fictive imagination, the reviewer implicitly buys into the language of Darwinian evolution by describing in his own words the features of the Martian creatures, using the analogy of familiar earthly fauna. Similarly, the *London Standard* described the Martians as something 'between a jellyfish and an octopus [...] Their round flabby bodies, covered with a shiny skin, are in fact hugely developed brains, life on the red planet being in a much more advanced condition than with us'.[81] Here, the concept of advancement of civilisations is described in biological terms, with the expectation that the development of life would implicate the transformation of bodies. A reviewer in *The Willesden Chronicle*, noting the 'law of the survival of the fittest', was even provoked to melancholic existential reflections, suggesting that 'this upstart planet' of Earth could yet be 'stillborn', and that 'we shall be squashed like [...] mites in the cheese'.[82]

While the London papers mostly focussed on the biological and other scientific aspects of *The War of the Worlds* in an initial flurry of reviews, regional papers tended to offer alternative accounts in the months following the book's publication. The *Lancashire Daily Post*, for example, covered a debate in letters between the Reverend Richard Fren of Millwall and Wells himself. Fren made 'a vigorous protest against one very grotesque creation' – not the hideous Martians, but the cowardly character of the curate in the novel. He accused Wells of either being ignorant about the English clergy, or playing 'to the gallery'. Wells responded that he was aiming to caricature 'an ignoble type, the curate of gentility', revealing a critical stance towards figures of religious authority. The columnist went on to discuss 'fashions' in the depiction of fictional archetypes, suggesting that Wells was writing for 'a generation that lusts after light literature'.[83] Implicit in the language of both the columnist and the correspondent is a kind of conservatism that was largely absent from the initial wave of reviews in the metropolitan newspapers. Speaking to such themes, the *Whitstable Times* chose to reproduce a short section of the novel that was described as 'a smart hit at the monotonous, humdrum city life'. Here, another of Wells' secondary characters, the artilleryman, outlines his vision of life under Martian rule, in which the suburban classes of London would happily submit to servitude, and come to 'wonder what people did before there were Martians to take care of them'.[84] Such commentaries on city life perhaps reflected a broader sense of the novelty of Wells' work, its metropolitan sensibility, and a greater sense of ambiguity in its reception outside of London. Here, a sense of geographical variation in the understanding of science-fiction in this period can be identified clearly.

Stapledon's two major works, *Last and First Men* and *Star Maker*, had an overwhelmingly positive critical reception across a broad range of journals, newspapers and periodicals, with reviewers often describing them in broadly superlative terms, as 'entirely original', 'a mad myth', or as 'deeply and widely imaginative'.[85] Commentators have also tended to elucidate a lack of conventional standards by which to measure Stapledon's writing, with one contemporary critic stating that *Star Maker* 'transcend[s] both novelistic writing and generic [science-fiction]: it is a new myth of creation and eschatology'.[86] While themes of divinity and creation were also common to the initial set of responses to *Star Maker*, such reactions were generally in contrast to that of the Christian and fantasy writer C S Lewis, who described the book as 'sheer devil worship', objecting to what he saw as Stapledon's clinical and irreligious portrayal of the cosmos.[87] Prominent in Lewis' mind is likely to have been the Star Maker of the book's title, a quasi-deity creator and destroyer of universes, with which the narrator comes face-to-face towards the end of the narrative. A number of other reviewers commented on the book's themes of eschatology and divine creation, considering Stapledon's sublime contemplation of the nature of the cosmos. In his review for *The London Mercury*, Bertrand Russell described the Star Maker as 'an artist rather than a philanthropist, [who] is perpetually increasing in ingenuity'.[88] For a reviewer in *The Illustrated London News*, 'the lord of the cosmos, if such there be, remains darkly aloof and inscrutable',[89] while *The Times* preferred the allusion of 'a God preoccupied, as the first and ultimate artist, with the creation of a perfection not perfectly foreseen'.[90] As such, themes of the sublime in *Star Maker* can be understood not only in relation to cosmic ascent and the contemplation of expanded scales of space and time, but also in a spiritual sense, as readers perceived a cosmic model of multiple universes and a dispassionate creator with no sense of benevolence to its progeny. Reviewers also noted the ways in which *Star Maker*'s cosmos conformed to modern scientific norms, 'always obeying the law of Gravitation (in the Einsteinian form of course)', with Stapledon himself characterised as 'an imaginative analogue of Einstein'.[91] Such references help in understanding the book's reception, in terms of the limitlessness of the sublime imagination, in parallel with the universality of the new laws of physics that had guided Stapledon's portrayal of the cosmos. They also put forward a way of understanding the cosmic realm as a modern myth, whereby spirituality and science are reconciled without recourse to the cosmologies of old.

Also common in reviews of *Star Maker* were themes pertaining to the more familiar scales of Earth and home, and allusions to the political context of Europe in the late 1930s. One reviewer saw Stapledon's final representation of the Earth as 'an arena where two antagonists were preparing for a critical combat', while another interpreted the book as 'a parable that clarifies the present crisis in our civilisation'.[92] Moreover, the *Liverpool Daily Post* recognised throughout *Star Maker* a critique of human society, including 'Nazism, Bolshevism, marriage, war, and so on', recognising a range of domestic and international tensions.[93] Most notably, the *Illustrated London News* drew attention to the book's final passages, where the protagonist returns home from his interstellar journey, 'to reflect upon the armed camps of the nations, the menace that may well annihilate all that is most precious to the individual'.[94] An often-quoted line from the book follows:

> Strange that in this light of the stars, in which the dearest love is frostily assessed, the human crisis does not lose but gains significance.[95]

What the reviewer takes from this is that humankind should 'confront the gathering storm with courage, to preserve moral integrity', suggesting the ways in which cosmic contemplation garnered meaning for individuals on the eve of the greatest conflict humanity had ever known.[96] Indeed, here is a sense in which the cosmic sublime could act as a source of comfort at times of uncertainty, not only through the experiences of the author, but also in the initial readings of *Star Maker*. With both Wells and Stapledon, therefore, it becomes apparent that readings of their texts derive meaning in different ways in different contexts, while the spatial aspects of such understandings, from metropolitanism to cosmographical reconfigurations, take on a renewed prominence.

Conclusion: foundational geographies of outer space

In a period before the science of space exploration became credible, even by marginal groups, literary engagements were greatly important in fostering understandings of outer space in Britain, which were conceived through a sense of cognitive estrangement experienced by authors and readers. Space, place and landscape had a central significance in the development of these understandings, exemplified in the work of H G Wells and Olaf Stapledon, from the laboratory spaces of South Kensington to the landscapes of the Welsh coast and the intimate writing spaces that acted to spur the imaginations of these two writers. Through this situated writing, Stapledon and Wells created expansive conceptual spaces of the fictive imagination, encapsulating eschatological alterities and implicating myriad variations of life across the Universe. The initial critical readership of their works has provided insights into the ways in which disruptive concepts in science and philosophy, including evolutionism, cosmic divinity and the rationality of the Universe, were made sense of in a variety of social and spatial contexts. In such spaces, the familiar became strange, and the strange became familiar, as understandings of outer space were conceptualised. As such, the foundational cultures and politics of outer space in Britain can speak to various contentions in human geography: that scientific knowledge is relative to social and cultural context, that place matters as much in the reception of ideas as in their generation, and that imaginative geographies offer a powerful means of synthesising expansive conceptions of space. Thinking through the fabrication of imagined and scientific understandings of outer space adds further dimensions to such considerations, incorporating additional aspects of spirituality, sublimity and scale to the geographies of science and society. Upcoming chapters consider the extent to which Wells and Stapledon helped to inspire and shape the theoretical and practical aspects to British space science and associated cultures, including the influence of Stapledon's thinking in later concepts of interstellar travel, and the impact of Wellsian themes on televisual representations of the alien 'other' in popular culture. Indeed, through both their lived experience of place and their imaginative geographies of science fiction, these authors changed the ways in which outer space was conceptualised in Britain for decades to come.

Notes

1   Robert Crossley, 'Famous Mythical Beasts: Olaf Stapledon and H G Wells', *Georgia Review*, 36 (1982), 619–635; Alan P R Gregory, *Science Fiction Theology: Beauty and the Transformation of the Sublime* (Waco, Texas: Baylor University Press) [Ch. 3: 'Wells and Stapledon – The Evolutionary Sublime']; Robert Shelton, 'The Mars-Begotten Men of Olaf Stapledon and H G Wells', *Science Fiction Studies,* 11 (1984), 1–14.
2   Darko Suvin, 'On the Poetics of the Science Fiction Genre', *College English,* 34 (1972), 372–382.
3   Angharad Saunders, 'Literary Geography: Reforging the Connections', *Progress in Human Geography,* 34 (2010), 436–452; Sheila Hones, 'Literary Geography: Setting and Narrative Space', *Social and Cultural Geography,* 12 (2011), 685–699; Sheila Hones, 'Literary Geographies, Past and Future', *Literary Geographies,* 1 (2015), 1–5.
4   Edward Said, *Orientalism* (London: Routledge, 1978); Bill Ashcroft, Gareth Griffiths, and Helen Tiffin, *The Empire Writes Back: Theory and Practice in Post-Colonial Literatures* (London: Routledge, 1989); Clive Barnett, '"A Choice of Nightmares": Narration and Desire in Heart of Darkness', *Gender, Place and Culture,* 3 (1996), 277–292.
5   Saunders, 'Literary Geography'.
6   Hones, 'Literary Geography'.
7   Rob Kitchin, and James Kneale, 'Science Fiction or Future Fact? Exploring Imaginative Geographies of the New Millennium', *Progress in Human Geography,* 25 (2001), 19–35; *Lost in Space – Geographies of Science Fiction*, ed. by Rob Kitchin and James Kneale (Trowbridge: Cromwell, 2002).
8   Kitchin and Kneale, 'Science fiction or future fact?' (p.4, p.2).
9   Suvin, 'On the Poetics of the Science Fiction Genre'.
10 Brian Aldiss, 'A Monster for All Seasons', in *Science Fiction Dialogues,* ed. by G Wolfe (Chicago: Academy Chicago, 1982), 9–24 (p.12).
11 Kitchin and Kneale, 'Science fiction or future fact?' (p.4) [emphasis in original].
12 Gale E Christianson, 'Kepler's Somnium: Science Fiction and the Renaissance Scientist', *Science Fiction Studies,* 3 (1976), 79–90.
13 Brian Stableford, *Scientific Romance in Britain, 1890–1950* (London: Fourth Estate, 1985).

14 Crossley, 'Famous Mythical Beasts' (p.622).
15 Stephen Daniels and Catherine Nash, 'Lifepaths: Geography and Biography', *Journal of Historical Geography,* 30 (2004), 449–458 (p.450).
16 David N Livingstone, *Putting Science in Its Place – Geographies of Scientific Knowledge* (Chicago: University of Chicago Press, 2003), p.183.
17 Herbert George Wells, *Experiment in Autobiography. Discoveries and Conclusions of a Very Ordinary Brain (since 1866)* (New York: J B Lippincott, 1967 [1934]), p.106.
18 Wells, *Experiment in Autobiography* (p.106).
19 Michael Sherborne, *H G Wells – Another Kind of Life* (London: Peter Owen, 2010), p.45.
20 Robert Crossley, *Olaf Stapledon – Speaking for the Future* (Liverpool: Liverpool University Press, 1994), p.32.
21 Crossley, *Olaf Stapledon* (p.2).
22 Michael Sherborne, *H G Wells.*
23 Crossley, *Olaf Stapledon* (p.171).
24 Robert Shelton, 'The Moral Philosophy of Olaf Stapledon', in *The Legacy of Olaf Stapledon – Critical Essays and an Unpublished Manuscript*, ed. by Patrick A McCarthy, Charles Elkins and Martin H Greenberg (London: Greenwood Press, 1989), pp. 5–22 (p.7).
25 Martin Danahay, 'Wells, Galton and Biopower: Breeding Human Animals', *Journal of Victorian Culture,* 17 (2012), 468–479 (p.471).
26 Wells, *Experiment in Autobiography* (p.192).
27 Sherborne, *H G Wells* (p.22).
28 Stapledon to Wells, 16 Oct 1931. In Robert Crossley, 'The Correspondence of Olaf Stapledon and H G Wells 1931–1942', in *Science Fiction Dialogues*, ed. by Gary Wolfe (Chicago: Academy Chicago, 1982), 27–57 (p.35).
29 Wells to Stapledon, 7 April 1936. In Crossley, 'The Correspondence of Olaf Stapledon and H G Wells' (p.39).
30 Crossley, 'Famous Mythical Beasts' (p.624).
31 Crossley, *Olaf Stapledon* (p.365).
32 Crossley, 'Famous Mythical Beasts' (p.635).
33 Robert Shelton, 'The Mars-Begotten Men of Olaf Stapledon and H G Wells', *Science Fiction Studies,* 11 (1984), 1–14 (p.11).
34 Wells, *Experiment in Autobiography* (p.458).
35 Herbert George Wells, *The War of the Worlds* (London: Penguin, 2005 [1898]), p.9.
36 K Maria D Lane, 'Mapping the Mars Canal Mania: Cartographic Projection and the Creation of a Popular Icon', *Imago Mundi,* 58 (2006), 198–211.
37 James R Moore, 'Revolution of the Space Invaders – Darwin and Wallace on the Geography of Life', in *Geography and Revolution*, ed. by David N Livingstone and Charles W J Withers (Chicago: University of Chicago Press, 2010), pp. 106–132.
38 George Levine, *Darwin and the Novelists: Patterns of Science in Victorian Fiction* (Chicago: University of Chicago Press, 1991); John Glendening, *The Evolutionary Imagination in Late-Victorian Novels: An Entangled Bank* (Aldershot: Ashgate, 2007).
39 Ranier Eisfeld, 'Projecting Landscapes of the Human Mind onto Another World: Changing Faces of an Imaginary Mars', in *Imagining Outer Space*, ed. by Alexander Geppert (London: Palgrave Macmillan, 2012), pp. 89–105; Christopher Keep, 'H G Wells and the End of the Body', *Victorian Review,* 23 (1997), 232–243; Danahay, 'Wells, Galton and Biopower'.
40 Ruth Barton, *The X Club – Power and Authority in Victorian Science* (Chicago: University of Chicago Press, 2018).

41   Adrian Desmond, 'Redefining the X Axis: "Professionals", "Amateurs" and the Making of Mid-Victorian Biology – a Progress Report', *Journal of the History of Biology,* 34 (2001), 3–50 (p.28).

42   Wells, *Experiment in Autobiography* (p.159).

43   Sophie Forgan and Graeme Gooday, 'Constructing South Kensington: The Buildings and Politics of T H Huxley's Working Environments', *The British Journal for the History of Science* 29 (1996), 435–468.

44   Wells, *Experiment in Autobiography* (p.160); Stableford, *Scientific Romance in Britain* (p.63).

45   Wells, *The War of the Worlds* (p.8).

46   Glendening, *The Evolutionary Imagination* (p.17).

47   Wells, *The War of the Worlds* (p.9).

48   Sherborne, *H G Wells* (p.143).

49   Herbert George Wells, *The First Men in the Moon* (London: Penguin, 2005 [1901]) (p.181).

50   Danahay, 'Wells, Galton and Biopower'.

51   Wells, *The First Men in the Moon* (p.182).

52   Crossley, *Olaf Stapledon* (p.174).

53   Crossley, *Olaf Stapledon* (p.176).

54   Saskia Warren, 'Audiencing James Turell's Skyspace: Encounters between Art and Audience at Yorkshire Sculpture Park', *cultural geographies,* 20 (2012), 83–102.

55   Crossley, *Olaf Stapledon* (p.174–6).

56   Saunders, 'Literary geography' (p.443).

57   Crossley, *Olaf Stapledon* (p.5).

58   Olaf Stapledon, *The Opening of the Eyes* (London: Methuen, 1954), p.29.

59   Crossley, *Olaf Stapledon* (p.186).

60   Philip Shaw, *The Sublime* (London: Routledge, 2006); Ralph O'Connor, 'From the Epic of Earth History to the Evolutionary Epic in Nineteenth-Century Britain', *Journal of Victorian Culture,* 14 (2009), 207–223.

61   Olaf Stapledon, *Last and First Men* (Harmondsworth: Penguin, 1987 [1930]), p.258.

62   Mariano M Rodriguez, 'From Stapledon's *Star Maker* to Cicero's Dream of Scipio: The Visionary Cosmic Voyage as a Speculative Genre', *Foundation – The Review of Science Fiction,* 118 (2014), 45–58 (p.47).

63   Olaf Stapledon, *Star Maker* (London: Magnum, 1979 [1937]), p.11.

64   Stapledon, *Star Maker* (p.12).

65   Stapledon, *Star Maker* (p.22).

66   Stapledon, *Star Maker* (p.105).

67   Stapledon, *Star Maker* (p.107).

68   Shelton, 'The Mars-Begotten Men' (p.11).

69   Stapledon, *Star Maker* (p.262).

70   Bertrand Russell, 'War in the Heavens', *The London Mercury,* 36 (1937), p.298.

71   Stableford, *Scientific Romance in Britain.*

72   Wells, *Experiment in Autobiography* (p.426).

73   Crossley, *Olaf Stapledon.*

74   Robert Crossley, 'Introduction', in *An Olaf Stapledon Reader,* ed. by Robert Crossley (New York: Syracuse University Press, 1997), pp.2–5.

75   Herbert George Wells, in *The Correspondence of H G Wells, Volume 1, 1880–1903,* ed. by David C Smith (London: Pickering and Chatto, 1998), p.300.

76   'The Scientific Novel – A Talk with Mr. H G Wells', *The Daily News,* 28th January 1898, p.6.

77   'Mars in Opposition', *Pall Mall Gazette,* 27th January 1898, p.4.

78 'The Library Table', *The Globe*, 24th January 1898, p.6.

79 'The War of the Worlds', *The Graphic*, 29th January 1898, p.148.

80 'Books of the Day', *The Morning Post*, 27th January 1898, p.2.

81 'Some New Novels', *The Standard*, 11th March 1898, p.8.

82 'Notes on News', *The Willesden Chronicle*, 27th January 1898, p.3.

83 'The Curate in Fiction', *Lancashire Daily Post*, 25th November 1898, n.p.

84 'A Satire on City Life', *Whitstable Times*, 19th March 1898, p.6.

85 Crossley, *Olaf Stapledon* (p.192, p.250).

86 Gerry Canavan, "A Dread Mystery, Compelling Adoration": Olaf Stapledon, *Star Maker*, and Totality', *Science Fiction Studies,* 43 (2016), 310–330 (p.323).

87 Clive Staples Lewis, in *From Narnia to a Space Odyssey – the War of Ideas between Arthur C Clarke and C S Lewis*, ed. by Ryder W Miller (New York: ibooks, 2003), p.40.

88 Russell, 'War in the Heavens', (p.297).

89 'Notes for the Novel-Reader: Fiction of the Month', *The Illustrated London News*, 31st July 1937, p.218.

90 'New Novels – Stars and Islands', *The Times*, 25th June 1937, n.p.

91 Russell, 'War in the Heavens' (p.297); 'A Philosophic Fantasy – Man and the Universe', *The Scotsman*, 15th July 1937, Books of the Day, p.15.

92 Roger Pippett, 'Galactic Utopia', *Daily Herald*, 24th June 1937, Books, p.14; Hugh Fausset, 'Adventures Among Stars and Men', *Yorkshire Post*, 30th June 1937, Among the New Books, p.8.

93 Walter Lyon Blease, 'A Cosmological Fantasy', *Liverpool Daily Post*, 4th August 1937, n.p.

94 'Notes for the Novel-Reader: Fiction of the Month', *Illustrated London News*, 31st July 1937, p.218.

95 Stapledon, *Star Maker* (p.).

96 'Notes for the Novel-Reader' (p.218).

3 Synthesising outer space

The British Interplanetary Society

The writings and readings of H G Wells and Olaf Stapledon were part of a broader upsurge in imaginative conceptions of spaceflight in the early decades of the twentieth century, as speculative thinkers in science fiction, and also popular science, considered the innumerable implications of leaving the Earth's physical boundaries. Indeed, ideas of cosmic transcendence had been circulating in Western culture since classical antiquity, including the imaginative possibilities of visiting other worlds.[1] While speculative narrations have continued to frame modern understandings of outer space, the early twentieth century also witnessed the development of new technologies that ultimately would turn the dream of cosmic ascent into the reality of spaceflight. The rocket itself has a long history in human society, from the Chinese 'fire arrows' of the middle ages, through its adoption as a ballistic missile in the Anglo-Mysore Wars of the late eighteenth century, and its subsequent deployment in the Napoleonic Wars by Sir William Congreve, to name just a few examples. However, the concept of using the rocket as the foundational technology of spaceflight was conceived independently in the early twentieth century by Robert Goddard (1882–1945) in America, Robert Esnault-Pelterie (1881–1957) in France, and Konstantin Tsiolkovskii (1857–1935) in Russia. They all published detailed works on spaceflight theory, understanding Newton's Third Law, which meant that a rocket would operate as a chemical form of propulsion in the vacuum of space. Other international figures, such as Hermann Oberth (1894–1989), Theodore von Kármán (1881–1963) and Frank Malina (1912–1981), took the rocketry principle a stage further with co-ordinated research programmes in Germany and the United States before the Second World War. As such, the age-old dream of transcendence became aligned conceptually with the technological potential of rocketry, a synthesis that was pursued by nascent spaceflight societies during the 1920s and 1930s, and which is the focus of this chapter with respect to the British Interplanetary Society.

The first of these societies was the *Verein für Raumschiffahrt* (Society for Spaceflight), established in 1927 by engineer Johannes Winkler in Breslau, Germany. This was followed by the American Interplanetary Society in 1930, set up in New York by the writer George Edward Pendray and science fiction editor David Lasser.[2] The British Interplanetary Society (BIS) was established in Liverpool in 1933 by Philip Ellaby Cleator, an amateur engineer and popular science writer. Other regional groups in the UK were known to have emerged around this time, including the Manchester Interplanetary Society and the Paisley Rocketeers' Society, which were merged into the BIS either side of the Second World War.[3] The BIS remains the longest-running spaceflight society in the world and while its pre-war years are the primary focus of this chapter, this book returns to the BIS in the post-war period, considering its substantial, ongoing role in the cultures and politics of outer space in Britain (see Chapters 4, 5 and 6). This chapter, however, deals with the ways in which the new science of 'astronautics' was established in Britain. This term originated in the works of the French science-fiction writer J-H Rosny *aîné* in the late 1920s to denote the science of space travel, and the fictive manner of the term's origins belies the reciprocity that occurred between science and the imagination in astronautics, especially, as this chapter argues, in the United Kingdom. More than just telling speculative stories about space travel, the BIS set out to demonstrate the concept of spaceflight as a tangible possibility within reach of human endeavour, while also taking the first practical steps necessary to make it a reality. In doing this across a variety of spaces in Liverpool and, later, London, the BIS enrolled a range of actors, materials and concepts, in a unique synthesis of interplanetary science and culture.

Historical geographies of outer space: synthesising technology and culture

Over recent decades, historical geographers have helped to demonstrate how scientific knowledge has been contingent on the vicissitudes of place, site, region and circulation. It is acknowledged that science 'takes place in highly specific venues; it shapes and is shaped by regional personality; it circles the globe in minds, on paper, as digitized data'.[4] Such contingencies have been found in examples ranging from ambitious programmes of international scientific co-operation, to the expeditionary practices of astronomical observation, and the intimate spaces of speech and conversation, in which scientific concepts have been formulated, tested and received.[5] Similarly, researchers in the history of science increasingly have focussed on understanding knowledge-creation in context, for example, in contrasting historiographies of iconic individual scientists, the importance of social relations among groups of scientific practitioners, or the significance of particular objects in the construction of scientific knowledge networks.[6] Drawing from work in social and critical theory, such accounts have questioned past perceptions of truth in science and society, while also attesting to the significance of contextual meaning and the subjectivity of knowledge.[7] Cognisant of such perspectives, a recent call for scholarly engagement with 'geographies of outer space' has highlighted the role of historical geography and the particular relevance of geographies of knowledge, geographical imaginations, and nature-society geographies, to explaining past understandings of outer space. As such, some of the historical aspects of space science, including the tracing of early maps of planetary bodies, the perceived composition of planetary environments, or 'the production of outer space knowledge claims' can be opened up to further geographical interpretations.[8]

Existing research in this area has interpreted understandings of outer space since the early modern era, demonstrating how geographers, critical of their own institutional histories of territorial mapping and enclosure, are well-equipped to critically appraise past representations of outer-space phenomena. As part of such enquiries, the representational and performative practices associated with the creation of nineteenth-century maps of Mars have been brought into question, problematising claims about the nature of planetary surfaces and highlighting the ways in which such 'truths' circulated among knowledge communities.[9] For example, astronomers who were seen to attain the best viewing positions of the night skies, often embarking on expeditions to remote sites in mountainous regions, were accorded a degree of legitimacy and prestige among their peers that stemmed from a sense of scientific masculinity, thereby qualifying processes of scientific observation with gendered notions of identity and mobility.[10] Historical geographies of the Space Age have further highlighted the situated significance of imagery, technology and perspective to the production of outer-space knowledges. Here, photographs of the Earth from space, taken by Apollo astronauts on the way to the Moon and in lunar orbit between 1968 and 1972, have been acknowledged for their iconic status in the history of environmentalism, while also being seen as emblematic of imperialism and global power by knowledge communities in different social and political contexts.[11] By contrast, the far lengthier perspective offered by the Voyager 1 space probe in 1990 facilitated a more elusive photograph of the Earth as a tiny speck in the vastness of space. This was Earth famously described as a 'pale blue dot' by the astronomer Carl Sagan, noting the paucity of human significance in grand cosmic narratives.[12] Such examples demonstrate how issues of power and representation have become central to understanding the ways in which outer-space knowledges have been produced, mediated and consumed across different spaces and contexts.

Further studies in historical geography and allied disciplines have recognised rocket technology as an essential aspect to the geopolitics of space science, including research that has highlighted the trans-Atlantic and post-colonial development of rocketry in the late modern era.[13] Pertinent to such understandings are the cultural and political foundations of rocketry in the early twentieth century, a time in which the concept of spaceflight was contingent on a more precarious mix of scientific cultures and untested technologies. In the United States, for example, the work of rocketry pioneer Frank Malina in setting up the Jet Propulsion Laboratory in 1930s

California has been recognised, as well as his subsequent replacement in hegemonic narratives of space exploration by the Nazi rocketeers who were naturalised as US citizens after the Second World War.[14] Here, actors in Cold War geopolitics, such as the infamous 'Un-American Activities' committee of the United States House of Representatives, played a prominent role in defining narratives of scientific achievement in outer space, and in the process conspired to obscure important earlier contributions. A similar story of historic achievement in space science that went unacknowledged for decades in official narratives is that of the Russian spaceflight theorist Konstantin Tsiolkovskii. Ignored by the state until the final years of his life, Tsiolkovskii fostered an active network of outer-space thinkers in early-twentieth-century Russia, exchanging correspondence with figures across Europe while teaching aeronautical theory in the district of Kaluga, near Moscow.[15] In such ways, research into the pre-history of spaceflight has highlighted the disconnect between official narratives of space exploration, such as American exceptionalism in space, or Soviet state leadership in spaceflight theory, and alternative accounts of individual and collective achievement. In both cases, the geographical specificities of place, on scales that were both local and global, played a significant role in the development and reception of scientific knowledge.

As well as illuminating the geographically situated foundations of space science, researching the origins of spaceflight can help to break down perceived divisions between amateur and professional scientists, as well as between popular and established science, concepts that have been challenged in recent years by scholars in the history of science.[16] Indeed, a limited number of studies have highlighted the significance of amateur societies, science-fiction fandom, correspondence networks and popular science writers in the early development of space science, not only in America and Russia, but also in Western European countries, such as Germany and the UK. For example, networks of outer-space enthusiasm have been identified in Weimar Germany, through which a combination of spectacular stunts involving rocket-powered cars, and media productions, such as Fritz Lang's science-fiction film *Frau im Mond* (1929), contributed to 'Germany's prominence in the early space travel movement'.[17] The emergence of the British Interplanetary Society in Liverpool around this time has also been conceptualised as a reflection of existing networks of science-fiction fandom and narration.[18] Looking at the broader European context, 'cosmopolitan networks of peripheral knowledge' have been associated with groups such as the *VfR* and the BIS in the mid twentieth century.[19] Here, the status of certain individuals as charismatic promoters of space science, located often at the fringes of mainstream science, is seen to be important in tying together emergent networks of knowledge, as is the international reach of such networks. Here is a sense in which space science emerged together with science-fictional and popular understandings of spaceflight in particular societal contexts, rather than through the institutions of traditional scientific practice. Such studies have also pointed the way in terms of a diversity of source materials in the historical geography of space science that go beyond iconic imagery and official documentation, towards recognising the social and cultural aspects of scientific practice and the interplay between imaginative and technological thinking in the making of new scientific territories beyond the Earth's surface.

The studies outlined in this section provide a context for considering the emergence of space science, or astronautics, in Britain before the Second World War. This work emphasises the importance of not just the social explanations of technology, nor simply the effects of technological development on society, but signify a hybrid approach that accounts for assemblages 'between diverse people, non-humans and places', while recognising such configurations as 'socio-technical imaginaries'.[20] Accordingly, this chapter builds on the previous chapter's research into the relationships between the fictive imagination, place and landscape, by focussing on the interplay between cultural, social and technological factors in the development of astronautics, factors which can be seen as uniquely important in the UK context. As such, the following sections explore the ways in which the British Interplanetary Society in its formative years embedded cultures of collaboration and internationalism within the emergent field of astronautics, while synthesising imaginative and practical approaches to space science. The emphasis turns firstly to the individual personalities and spaces of science that were central to the establishment of the BIS, then to the ways in which the Society forged international connections and the significance of this, and finally considering its work on conceptualising space exploration through the design of a lunar spaceship.

Establishing astronautics: P E Cleator, A M Low, and the British Interplanetary Society

In 1930s Britain, astronautics was not an established field of science in the sense of being a feature of educational curricula, researched in government scientific institutions, or developed by large engineering corporations. At this time, the science of aeronautics, which included 'heavier-than-air' conventional flight as well as 'lighter-than-air' balloons and airships, had itself been in existence for just a short period, while rocketry consisted only of small self-propelled projectiles used for fireworks and sometimes as short-range ballistic weapons. While individual pioneers and experimental research groups in other countries such as had started to theorise about astronautics, and in some cases progressed to the testing of high-altitude rockets, their work was either ignored by governments (in the case of Tsiolkovskii and Goddard) or manoeuvred into ballistic or aeronautical weapons research (in the case of Oberth, von Kármán and Malina). In Britain in the early twentieth century, there were no abstract theorisers of spaceflight, let alone practical experimenters, partly because the Explosives Act of 1875, passed in the context of domestic unrest posed by advocates of Irish Home Rule, meant that any applied research with rockets was effectively prohibited. Yet, importantly, there remained a strong culture of speculative writing about spaceflight in Britain, evidenced through the works of Stapledon, Wells, and others in 'scientific romance' fiction (see Chapter 2), while an emerging popular science literature had also started to consider some of the practical possibilities of spaceflight.[21] It was inevitable, therefore, in the increasingly globally connected societies of the early twentieth century, that some of the new theories of spaceflight would become known to interested individuals in Britain, and that the concept of space exploration would take hold in the imaginations of many others. The main way in which these ideas formulated in Britain was through the British Interplanetary Society, and considering the ways in which this group coalesced in Liverpool in the early 1930s, and London towards the end of that decade, can illustrate how technologies and cultures of spaceflight were closely integrated in the emergence of astronautics in Britain.

The BIS was founded in 1933 by Philip Ellaby Cleator (1908–1994), an amateur scientist and popular science writer from Wallasey, near Liverpool. Initially, the Society attracted a handful of young members from the Liverpool area, who were overwhelmingly male and came from a variety of middle- and working-class professions including engineering, journalism and publishing. Its membership rose gradually to over a hundred, and among this initial group, there was a strong interest in science-fiction, with some members affiliated with a Liverpool-based science-fiction club called 'The Universal Science Circle'.[22] The growth of popular science publications during this period also provided an opportunity for the Society to circulate its ideas, with, for example, notifications about BIS meetings being published in *Practical Mechanics* from 1933.[23] The BIS was notable for the publication of its own regular journal, which was established in 1934 and continues today as the world's longest-running journal on astronautics. The contents of the *BIS Journal* reflected the broad basis of interplanetary science as conceived by the Society, with articles published on a diverse array of topics, from the existence of alien life in outer space, to the latest developments in rocket technology and concepts of interplanetary navigation. Indeed, while it started out as a local group, the BIS always had broader ambitions, both in the sense of extending its membership to national and international scales, but also in terms of its intellectual activities, thinking beyond just rocketry to theorise about the myriad possibilities of interplanetary space. This ambition led to the Society transferring its headquarters to London in 1936, where its membership was growing more rapidly and was better connected to influential figures, such as the Society's second President, Archibald Montgomery Low (1888–1956). Further examination of the Society's leadership figures will help to outline the ways in which the social and the technical aspects of space science became co-reliant in the formative years of the BIS.

Philip Cleator had a life-long fascination with the concept of spaceflight, and his overlapping enthusiasm for science-fiction, popular science and outer space led to him becoming the driving force behind the BIS in its Liverpool years. A committed writer as well as an amateur scientist, Cleator had published a number of fictional and non-fiction works on interplanetary themes by the mid-1930s. Typical of this writing was a short article titled 'The Possibilities of Interplanetary Travel', which outlined what Cleator knew of rocket research being undertaken around the world and the potential application of this work to spaceflight. It ended with the conviction that, 'whether we like or dislike the idea, believe or disbelieve, interplanetary travel will come.'[24] The release of his first book, *Rockets Through Space, Or, The Dawn of Interplanetary Travel*, consolidated Cleator's legacy as one of the first people in Britain to theorise and publish substantial works on space science.[25] In September 1933, Cleator had a letter published in the *Liverpool Echo* as a call for interest to form an interplanetary society.[26] While not having much of an immediate impact in terms of prospective new members, the letter drew the attention of N E Moore Raymond, a correspondent of *The Daily Express* newspaper, who subsequently arranged to meet Cleator in person. Following this encounter, Cleator was able to secure a front-page article in the *Express* advocating the idea of establishing an interplanetary society, after which its first would-be members began to contact him.[27] It was, therefore, Cleator's ability as a writer, mediator and publiciser that had helped him establish the BIS in the first instance, while the scientific concepts of interplanetary travel had largely been already established by this time.

Space historian Frank Winter, who interviewed Cleator in 1978, suggested that it was his 'maverick individualism' that helps to explain Cleator's role as a champion of interplanetary science in this period.[28] This was a time in which the notion of space travel received hardly any broader credibility as a serious scientific endeavour and it could well be argued that it was Cleator's sheer force of personality that convinced many people in Britain of the potential of spaceflight. He was profiled widely in the national and international press upon the release of *Rockets Through Space* in 1936, with the *Yorkshire Post* describing him as 'a visionary' in an illustrated full-page article and the *New York Post* announcing him as a 'well-known scientist', with reviews praising the book for its detailed reporting of the latest developments in rocketry and its wide-ranging speculations on concepts of spaceflight.[29] Cleator's interest in language and the written word was reflected in his decades-long correspondence with the prominent American journalist and critic Henry L Mencken (1880–1956), in which they discussed astronautics, as well as their shared interest in aspects of culture and politics.[30] Cleator's command of the English language, as well as his sometimes acerbic wit, undoubtedly helped him to communicate the science of outer space effectively, while his public profile as a type of charismatic visionary scientist certainly helped to publicise the interplanetary concept to a wide audience. Through this profile, he perhaps embodied an early version of the archetypal figure of the 'rocket scientist', that featured prominently in post-war popular representations of science.[31] The peak of Cleator's involvement with the BIS came around the time that his book was published, and as its centre of gravity was transferred to London in 1936, his influence in the Society he founded started to wane.

The second President of the BIS, Archibald Montgomery Low (pictured centrally in Fig. 3.1), was a freelance radio engineer, inventor and author, and has been presented in his biography as a flawed genius who 'never reached the heights to which he had a right'.[32] His inventions included an early form of television in 1914 that he called the 'TeleVista', and he had patented well over a hundred devices, in areas such as radio guidance, infra-red photography and internal combustion engines. He had already been awarded Honorary Membership of the BIS during the Liverpool era and his status as a kind of arbiter between the worlds of the professional and the amateur seemed to be a major attraction to the BIS, as it was looking for a figurehead for its new London headquarters. At the same time, Low found himself the subject of 'bitter antagonism' and 'much personal abuse and ridicule' from some professional scientists because of his somewhat ambiguous academic standing.[33] Although he held a position as 'Associated Honorary Assistant Professor of Physics' at the Royal Ordnance College from 1919 to 1922, Low continued to trade on his status as 'professor', even though he was not strictly entitled to do so.[34] This controversy is said to have indicated 'a tension between the elite and those who focussed exclusively on

INAUGURAL MEETING IN LONDON, OCTOBER 27, 1936.

Back row (left to right): Unknown, C. G. Smith, Allen, *J. G. Strong, Unknown, *R. A. Smith, Unknown, F. Day, *C. Bein, *M. K. Hanson, Unknown.

Front row: Unknown, E. J. Carnell, *A. C. Clarke, *W. H. Gillings, *A. M. Low, Dubois, J. H. Edwards, Miss E. Huggett.

* Present members of the Society.

Figure 3.1 First London meeting of the BIS in October 1936

Source Credit: British Interplanetary Society

applied science'.[35] This did not seem to matter to members of the BIS, and Cleator, for one, viewed Low as 'a brilliant eccentric with a penchant for embracing what, at the time, were dismissed as wild and improbable ideas', while post-war BIS Chairman Val Cleaver described Low as an 'interesting personality [who] has a flair for publicity which might be useful'.[36] Low's connections in the publishing industry were to prove valuable, and as editor of the popular science magazine *Armchair Science*, he commissioned articles from BIS members on various interplanetary topics. Low is known also to have penned a number of science-fiction stories, including such titles as *Adrift in the Stratosphere,* and he believed that speculative fiction could, in fact, encourage real developments in technology that would be of great benefit to society.[37]

In these respects, Cleator and Low had much in common. Both were seemingly aware of the effect of their public profile and the need to forge connections between the emerging science of astronautics and the public-at-large, through embracing the printed media, especially science-fiction and popular science. To a certain extent, Cleator and Low conform to the figure of the 'space persona', being 'a particular type of scientific person [...] presupposing a certain degree of socio-cultural recognition', an archetype whose importance has been recognised in shaping the pre-history of space science, not just in Britain but across Europe.[38] Certainly, examining their respective circles of contacts and engagements helps to emphasise the value of social relations and popular interest in supporting the early development of astronautics in Britain.

Alternative spaces of science: the British Interplanetary Society in Liverpool and London

While the BIS relied on its charismatic figureheads to outwardly promote the Society to potential new members and audiences, another significant aspect to the Society's functioning in its early years concerned the spaces in which it operated. At this time, the only sources of income for the BIS were membership subscriptions and small donations, which were said to have 'barely covered running expenses', such as the production of the flagship *BIS Journal,* and it had no formal connections with public or scientific institutions.[39] Consequently, the Society made use of a variety of public and private locations during its formative years, in which members held meetings, socialised and exchanged information. Studies in the history of science have recognised the importance of diverse spaces of scientific activity, including pubs, tea rooms and coffee houses, in understanding the practice, discussion and dissemination of scientific knowledge since the early modern era.[40] Accordingly, 'the perspective gained from focussing on the place of science', especially places such as the public house, can further enable the understanding of science as a group practice.[41] In relation to this, there has been a growing recognition of the importance of 'geographies of oral communication' including oratory, conversation and other communicative performances, in the functioning of scientific knowledge communities.[42] In such ways, the dynamics that occur between space and sociality have been recognised as important aspects of the scientific cultures of the past, and looking at some of the spaces of the BIS, therefore, becomes important in understanding the emergence of astronautics in Britain.

One of the key spaces of science during the Liverpool era of the BIS was Philip Cleator's house at 34 Oarside Drive, Wallasey, at which a variety of social and scientific practices took place.[43] This was also the headquarters of Cleator's engineering business, the 'Science Research Syndicate', which he had taken over from his parents, along with substantial laboratory equipment and materials. Here, the first informal meeting of the BIS was convened in September 1933, at which the five attendees, Cleator, Leslie Johnson, Colin Askham, Herbert C Binns and Norman Weedall, 'solemnly pledged [them]selves to begin the monumental task of making the inhabitants of Great Britain rocket-conscious and astronautically minded'.[44] Johnson later recalled this encounter, indicating the tone of the meeting and detailing some of the activities that took place:

> [A]bout half-a-dozen enthusiasts turned up that evening, when they were entertained at Philip Cleator's laboratory, shown an embryo rocket motor, and had demonstrated to them the unstable explosive qualities of Fulminate of Mercury.[45]

In this initial gathering, Cleator's laboratory was used as a space of demonstration, display and entertainment. It is unlikely that Cleator's 'embryo rocket motor' was very sophisticated, but having practical props such as this would certainly have helped to demonstrate the concept of rocket propulsion to his eager audience. In experimenting with Fulminate of Mercury, Cleator was, probably unknowingly, in breach of the aforementioned Explosives Act, yet the demonstration showed that he understood the general principle of rocketry as based on the containment of an explosive chemical reaction, and would also, no doubt, have added a sense of theatricality to the occasion. Here, the kind of group practice that took place can be understood in terms of performance, occurring through the interplay between the scientific practitioner and the audience, along with the use of props. In this way, the domestic space of Cleator's home was turned over effectively to a scientific space, and a meeting place for the fledgling Society. It is also worth noting that, at the same time that Cleator was hosting his guests in Wallasey, Olaf Stapledon was living under ten miles away in West Kirby, on the other side of the Wirral Peninsula, part of the suburban locale of Liverpool, at a time when his writing career was entering its most productive phase. While the two never knowingly crossed paths (Stapledon was later to correspond with the BIS after the war), it is notable how, in both cases, suburban domestic spaces became instrumental in speculative thinking about outer space.

Furthermore, proximity to the globally connected city of Liverpool may have had a common influence on both Stapledon and Cleator through a shared exposure to internationalist cultures and politics.

As well as Cleator's house, BIS meetings were held in a variety of other locations in and around Liverpool in the early 1930s. Different spaces served different purposes, and in October 1933, the Society's inaugural meeting occurred in the office of H C Binns at 81 Dale Street, Liverpool. This was said to be 'typical of a solicitor's or accountant's office [...] with dark woodwork and translucent glass panels', and the official setting served to formalise the meeting, which unanimously resolved that the British Interplanetary Society be formed, while electing the roles of President (Cleator), Vice-President (Askham) and Secretary (Johnson).[46] Over time, the Hamilton Café, at 56 Whitechapel, Liverpool, became the Society's regular meeting-place, which was later described by Cleator as,

> [A]n unpretentious place, which exactly suited our needs: it was centrally located; it remained open until late hours, it provided light refreshments as and when required; and its charges were absurdly cheap.[47]

Meetings at the Hamilton Café typically occurred in the evenings, covering administrative matters related to the Society, as well as holding 'Free Discussions' on topics such as 'The Temperature of an Object in Space' and 'Can We Plot a Path for a Space-Ship from the Earth to One of the Planets?'[48] One meeting in February 1936 featured the visiting German rocket experimenter and *VfR* member Willy Ley. Ley was reportedly on his way to America, described as 'a fugitive from Hitler's Third Reich'.[49] This meeting attracted a number of local newspaper reporters, at which Ley gave an address to the dozen or so members of the Society who were present, reportedly on the subject of sending a rocket to the Moon. A bespectacled Ley was subsequently pictured in *The Daily Sketch* showing some plans to surrounding members of the BIS, including Cleator and Askham, which were later pointed out to have been nothing more than 'a blank sheet of foolscap paper', used as a staging prop to convey a sense of scientific realism.[50] This episode highlights the role of hospitality and the staging of media events in the enactment of interplanetary science during this period. In this way, the practices of group scientific activity were closely aligned to the various types of place in which the BIS met, as well as to their broader international networks, enabling a diverse range of activities and modes of engagement.

A similar variety of places were utilised by the BIS once it had settled down in London by 1937, at which point a new cohort including Arthur C Clarke (1917–2008), Harry Ross (1904–1978) and Arthur 'Val' Cleaver (1917–1977) had become more prominent in the Society (Clarke is pictured in Fig. 3.1 to the left). Here, members are said to have met 'at least once a week in cafés, pubs, or each other's modest apartments', the most frequented locations being the Mason's Arms and the Duke of York, two pubs in the vicinity of London's Oxford Circus.[51] In a retrospective account, Ross recalled how 'pre-war members of the BIS have met in many strange places – some of them palatial, some of them markedly otherwise, most of them pubs', remembering one occasion in 1936 featuring 'a sea of about 30 fantastically assorted faces, an incredible aroma of fish, beer and tobacco, a glass-fronted wall-case of toy soldiers, and a heck of a noise'.[52] Val Cleaver similarly recalled discussions at the Duke of York in 1938, one on the subject of a 'proposed launching device for a spaceship' that would float in a sea-borne tank, and another at which a visiting member of the American Rocket Society, Midshipman Robert Traux, displayed a prototype fuel-cooled rocket motor, the first Cleaver had ever seen, and 'told us some very tall American stories'.[53]

These accounts highlight some of the social interactions within the BIS, the atmospheres in which they took place, and their role in the formulation of astronautics, with exchanges of ideas, the telling of stories and the display of objects constituting a large part of the highly speculative endeavour that was 1930s' space science. Such interactions in London echo the kinds of activity held previously in Liverpool, where BIS members from modest backgrounds, with little in the way of conventional training, conducted science in informal settings, focussing on the fundamental problem of achieving spaceflight. While their practical experimental work was negligible in comparison to early rocketry programmes occurring around the same time in Germany and the United States, the Society undertook vigorously the broad intellectual labour of spaceflight, including theoretical work, discursive debate, public promotion and material circulation of astronautical knowledge. Looking at the social and spatial contingencies of this work, it appears that, far from being hindered by a lack of official premises in which to conduct their business, the diversity of informal spaces that were used by the BIS effectively enabled the development of astronautics amid a vibrant international scientific culture.

Networking and theorising internationalism in space science

Despite its status as a national society, as implied by its name, and the inevitable geographical restrictions of such an organisation, a strong feature of the British Interplanetary Society since its foundation has been its sense of internationalism, not just as a practical consideration but also as a cultural and political outlook. Studies in historical geography have sought recently to interrogate 'the international' in science, culture and politics with a particular focus on the early-and mid-twentieth century.[54] While tracing some of the global networks of internationalism, this work has also looked to define its more 'aspirational' qualities, including the ways in which 'science and scholarly research sought to reformulate [themselves] as forms of intellectual practice' in light of the ideals of internationalism.[55] This can be seen in discourses of airship science and technology during the early twentieth century, with the Earth's atmosphere being seen as 'the natural domain of internationalism' by airship promoters who sought to bring together cultures of empire, atmospheric science and international mobility.[56] Similar qualities can also be seen in the development of astronautics, with individuals in America, Russia and Europe forming theories about astronautics independently, then to varying degrees engaging and collaborating with a growing international community of spaceflight theorists. Yet astronautics, in its purest and most ideological sense, was a science orientated around the possibility of transcendence, both in terms of escaping the confines of the Earth's gravity, but also, in doing so, rejecting Earth-bound notions of territory and the divisions of nation-states. Arthur C Clarke would later become one of the most eloquent proponents of this theoretical side of astronautics, stating in 1946 that 'it is not easy to see how the more extreme forms of nationalism can long survive when men begin to see the Earth in its true perspective as a single small globe among the stars'.[57] This emergent philosophy, in some ways an extension of earlier thinking that helped to de-centre the Earth from universal models of the cosmos, became one of the principal reasons why the BIS sought to embrace internationalism in the intellectual practice of astronautics, over and above the practical necessities of collaboration across borders.

These multifaceted qualities of internationalism were evident in the early years of the BIS, and Philip Cleator's very first influences in astronautics were wholly international, from his initial correspondence with members of the American Interplanetary Society in New York, to the news reports he read about 'the experimental activities of the *VfR*' in Germany.[58] Liverpool's active science-fiction cultures, in which many early BIS members participated, also had an international basis rooted in the city's imperial maritime heritage.[59] During the early twentieth century, as science-fiction print culture was established in New York through the work of editors such as Hugo Gernsback and John Campbell, 'pulp' magazines such as *Amazing Stories* and *Astounding* were routinely transported to the UK as ships' ballast, where they were sold on by the Woolworths chain of newsagents.[60] As well as reaching members of the BIS in Liverpool, the American pulps were a major influence on the scientific imagination of Arthur C Clarke, who later wrote his nostalgic 'science fictional autobiography' as a paean to these often wildly imaginative magazines in the 'Golden Age' of American science fiction.[61] Through such avenues, British writers, such as Clarke and William F Temple, another early BIS member, published their first science-fiction stories, and their influence was also seen in the fleeting appearance of the first British science-fiction serial, *Scoops*, which ran to twenty issues in 1934.[62] As well as introducing fantastic narratives of space exploration, science-fiction editors also encouraged interaction among their readership, primarily through special correspondence pages, which naturally reflected the international reach of their circulation. Leslie Johnson recalled that one such title, *Wonder Stories Quarterly*, had carried a letter in its Summer 1930 edition announcing the formation of the American Interplanetary Society, and it was through correspondence with this group that Cleator was first encouraged to set up the BIS, establishing a connection that continued to be maintained thereafter.[63] In this way, the international circulation of science-fiction publications helped to inspire the formation of the space-flight societies, and acted as a model for the exchange of their own publications in the years to come.

Within its first few editions, the *BIS Journal* was reporting ongoing developments in the overseas societies, featured under the title 'International Interplanetary News'. Here, updates from around the world were reported, including the construction of test rockets by the American Interplanetary Society in New York, the delivery of a speech in Austria on the subject of astronautics by the rocketry experimenter Guido von Pirquet, and the confirmation of science writer Iakov Perelman of Leningrad as a Fellow of the BIS, amid other goings-on.[64] Furthermore, the *BIS Journal* began to publish articles by overseas experts in astronautics. In one early issue, a paper by the German experimenter Otto Steinitz theorised some of the physical processes involved in launching a rocket into space, noting accurately that 'as a result of fuel consumption, the weight of the ship will continually decrease, while its momentum will progressively increase', an important factor in calculating the relationship between fuel loading and orbital escape velocities.[65] Contributions from other overseas writers followed, including Willy Ley, G Edward Pendray and Werner Brügel. By developing an international network of contacts in this way, the BIS was able to draw upon expertise from the German and American spaceflight societies, as well as freely reporting the results of these groups' practical experiments.

Through such reporting and correspondence, the BIS had started to engage internationally, yet there were hints of a more substantial, intellectual commitment to internationalism. In *Rockets Through Space*, Cleator mentions a proposed 'International Commission for Astronautics', while promoting open information exchange between all the world's rocket societies.[66] He believed that this approach would reap benefits for the science of astronautics, not only in helping to avoid redundant lines of enquiry and associated duplication of effort, but, more fundamentally, in paving the way for 'an international organisation, free from national bias, whose personnel, approved by the world, might be ultimately entrusted with the funds and power necessary to build the first space-ship'.[67] In such ways, Cleator and other members of the BIS took the concept of internationalism beyond the practicalities of collaboration, towards theorising more fully about the ideals of internationalism in science and what this could mean for astronautics.

Furthering such ideas, Cleator arranged a visit to the renowned Berlin-based rocket society, the *Verein für Raumschiffahrt* (*VfR*), in early 1934, which would be instrumental in formulating the Society's internationalist outlook. This group attracted an international membership of engineers, scientists and laypersons, and was representative of a 'space fad' that had taken hold in parts of Europe and Russia in the 1920s.[68] Unfortunately, for Cleator, by the time of his visit the *VfR* was defunct, essentially taken over by the German Army, but in spite of this, he was able to meet his contact, Willy Ley, and on his return, wrote enthusiastically about his visit in the *BIS Journal*:

> Thanks to Herr Ley, I was able to obtain introductions to many of the leading experimenters throughout the world [...] Such generosity of action [...] exemplifies the true international nature of the scientific spirit.[69]

Cleator's visit to Germany can thus be seen as a kind of small-scale diplomatic mission, whereby cordialities and information were exchanged for mutual benefit and the advancement of spaceflight research. Indeed, the visit became a formative experience in Cleator's outlook on spaceflight research, convincing him of the necessity of international co-operation. This was reflected in the policies of the BIS, as it started to establish connections across the world, including societies in America, Italy, France, Austria and the USSR. As part of such efforts, the *BIS Journal* reported attempts to 'establish radio communication between members of the world's rocket societies', while an international journal exchange programme was also initiated.[70] The publications involved in this scheme were the *VfR*'s *Die Rakete*, another German publication *Das Neue Fahrzeug*, for which Cleator was to pen an article himself, the *Bulletin* and *Astronautics* of the American Rocket Society and *Space* of the Cleveland Rocket Society, with the aim of establishing 'a free exchange of vital information between the rocket societies of the world'.[71]

Perhaps concerned about the problem of translation in international collaboration, the BIS even took 'a paternal interest in Ido (reformed Esperanto) as an international language', and, as a result, an article by Leslie Johnson entitled 'Space Rockets' or '*Space-Fuseli*' was published in *The Ido Magazine*, while an advertisement for '*Centerbladet* – a journal with a special appeal for all interested in international language problems' appeared in the *Journal of the BIS*, with the Society also receiving copies of the *Ido–English Dictionary* and *Montala Letro*, both published by the Ido Society of Great Britain.[72] By turning to an international language, the BIS tapped into a wider cultural and political project that sought to reject 'emerging nationalisms and linguistic chauvinisms'.[73] As such, the new languages such as Ido and Esperanto were regarded as solutions to the geopolitical tensions of the late-imperial era that would entail further benefits for global cultural and scientific development in the twentieth century. This provides an intriguing example of how the BIS used innovative, optimistic thinking in coming to terms with the 'abstractions of internationalism', as well as dealing with the practical implications of international collaboration.[74] Such conceptual challenges, in the context of the Society's changing membership and varying capabilities, were reflected in the technical designs of the pre-war Society, which considered some of the fundamental issues related to the prospect of human spaceflight.

Conceptualising space exploration: the BIS space-ship

Understanding the context in which the BIS emerged and undertook its work has helped to demonstrate the unique circumstances in which astronautics was established in Britain. As well as creating foundational knowledge networks and setting in place some of the basic principles of astronautics, the BIS also advanced several concepts of space exploration that foreshadowed the advent of the Space Age. In contrast to both the German and American spaceflight societies, which focussed on practical rocket testing and made important contributions in areas such as propulsion and aerodynamics, the BIS turned its attention to the bigger picture of space exploration, whilst still being able to draw influence from the work of these other groups.[75] Re-affirming and explaining the desire for space exploration was, therefore, one of the priorities of the BIS as it completed its organisational shift to London in 1937. Addressing this issue, A M Low, writing his first address as President, characterised BIS members as 'pioneers' who 'know perfectly well that spaceships are a matter of time and patient research'.[76] The immediate task of the BIS, therefore, was to 'influence public opinion' and 'explain that we are not peculiar people who desire to go to the Moon like children who cry for a new toy', in elucidating its carefully laid plans in a credible and believable manner.[77] This somewhat defensive rationale perhaps grew out of Low's experience at the fringes of the scientific establishment, and the wider opprobrium that he knew astronautics attracted at this time. By contrast, William F Temple, the science-fiction writer and new editor of the *BIS Journal*, professed more lofty motivations, stating that 'the BIS had chosen exploration, to help in the work of pushing the boundaries of known territory as far as we can, sheer across the universe if possible'.[78] Tellingly, he quoted Olaf Stapledon, whose hero in the novel *Last Men in London* (1932) declared 'an obscure impulse to devote himself to ends beyond private gratification', his mind having been infiltrated by one of Stapledon's omniscient 'last men' of the planet Neptune.[79] As such, the desire for space exploration crystallised from a variety of impulses, from science-fiction-inflected optimism to the premise of expanding humankind's territorial reach, centring around the notion of the 'greater good' of scientific endeavour.

Spurred on by such notions, in 1939, the BIS produced a single design concept that synthesised the cumulative knowledge of the Society, including Cleator's book *Rockets Through Space*, existing works in the *BIS Journal*, and reports from German and American spaceflight societies, bringing together the dual challenges of rocket technology and the science of human existence in outer space. In the words of R A Smith, an engineering drafts-man and one of the key members of the BIS at this time,

> The case under consideration is to show how a vessel may be con-structed to carry a crew of two safely to the Moon; permit of their landing for a stay of fourteen days; [and] provide for their safe return with a payload of half a ton.[80]

This project became known as the 'BIS space-ship' and was carried out by a 'Technical Committee' consisting of Smith, H E Ross, M K Hanson, J H Edwards, A Janser, H Bramhill and A C Clarke. This team of researchers were described by Temple as 'practical idealists', characterised by a rare combination of technical thinkers, 'who believe in the future of the rocket motor', and imaginative thinkers, 'who believe that one day man will cross outer space to the planets'.[81] This combination was an essential character-istic of the BIS at this time, and grew out of a context in which the Society was prevented, both legally and financially, from conducting extensive tech-nical experiments, but fostered instead an intellectual environment in which the imagination could thrive in tandem with the promise of science and technology.[82] With a broader range of active and enthusiastic committee members in London, who were able to work together as a focussed team, the Society was able to produce more detailed design concepts than in the Liverpool era, and their design was indeed the first comprehensive study of the feasibility of a manned mission to the Moon. The Technical Committee met at least once a month, at informal central London locations, producing a series of detailed reports for the *BIS Journal* that outlined several aspects of the BIS space-ship, including motor design, payload specifications and navigation in space.

Imagery showing details of the BIS space-ship adorned the front cover of the *BIS Journal* in January 1939 (Fig. 3.2). Here, cut-away diagrams revealed the cellular design of the propulsion system, 'with hundreds of small solid-propellant units each providing thrust individually, and each so attached that as soon as it ceases to thrust, it falls off'.[83] This was based on the principle of staged rocket launches, as previously outlined by Cleator and others, and was recognised as one of the most effective ways in which the efficiency of a space rocket could be optimised. The Technical Committee based its calculations on existing work in Europe and elsewhere, includ-ing by Hermann Oberth and Eugene Sanger, both formerly of the *VfR*, as well as Robert Goddard in the United States. It also drew its conclusions from critical analysis of the rocket-car tests of Fritz van Opel in Germany in the late 1920s, which used liquid propellants to varying degrees of suc-cess in a series of widely reported stunts. Thus, a cellular design that used solid fuel was arrived at, being 'an attempt to incorporate the advantage of the rocket car type of design in conjunction with methods of avoiding its greatest disadvantages [and] in conjunction also with the principles of

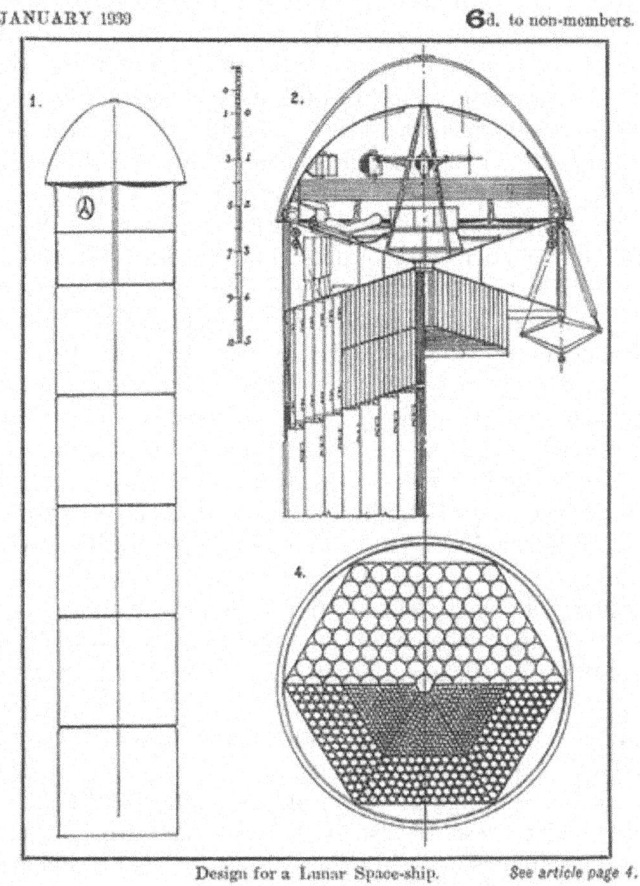

JOURNAL
OF THE
British Interplanetary Society

JANUARY 1939   **6**d. to non-members.

Design for a Lunar Space-ship.   *See article page 4.*

Figure 3.2 Front cover of the BIS Journal displaying diagrams of the BIS space-ship

Source Credit: British Interplanetary Society

step construction.'[84] As well as the main thrusters, the BIS space-ship also incorporated a number of motors to manoeuvre the rocket through space, including rotating the ship through 180 degrees to prepare for landing on the Moon, and using jets of 'steam at high pressure' for the more delicate adjustments during the voyage.[85]

The interlinked problems of gravity and navigation were also addressed. To this end, a number of manual instruments were listed as part of the proposed inventory, including an altimeter, a speedometer, a gyroscope, a chronometer, and 'space-sextants' for observing the positions of heavenly bodies as the space-ship made its way towards the Moon.[86] In this way, the spaceship would be steered actively by its crew towards the Moon, rather than relying purely on pre-ordained trajectories. However, the task of navigation was made more complicated by the fact that the space-ship was set to spin around its vertical axis at a rate of once every three-and-a-half seconds. This was deemed necessary, both to stabilise the rocket and in order to create an artificial gravity, as the effects of prolonged exposure of astronauts to a zero-gravity environment were a cause of uncertainty, with some anticipating a permanent state of vertigo. Due to this constant rotation, any accurate viewings of the background stars from the ports in the astronaut compartment would be nearly impossible. A solution to this problem was devised through the use of a 'coelostat', an optical device that uses motors and mirrors to neutralise the visual distortion caused by the ship's rotation. A proof-of-concept prototype was constructed by the technical committee using an old gramophone motor, which was demonstrated at a BIS meeting, in what was the only experimental device constructed by the Society at this time.[87]

The 'lunar spaceship' diagram also reveals a human compartment (Fig. 3.2). This would provide enough space for three crew members and all the necessary supplies to maintain them for the duration of the journey, including a fortnight on the Moon itself. Supplies of hydrogen peroxide would be used as a source of water and oxygen, while food supplies would mostly consist of high-calorie sustenance including 'bread and butter, cheese, porridge, chocolate and sweet cocoa', with a small provision of coffee, salt and even an alcoholic beverage 'to celebrate the landing on the Moon'.[88] To survive excursions to the lunar surface, the astronauts would wear space-suits made of rubber or leather, with integrated heating and a supply of oxygen.[89] Some understanding of the expected conditions on the Moon were conveyed, with the dangers of quick-sand-like dust as well as sharp-edged lunar rocks accounted for in the design of the excursion equipment. Research on the Moon would involve telescopic astronomical observations, the collection of mineralogical specimens, as well as searching for 'spores, lichens, or other forms of life'.[90] Such expectations revealed that BIS members saw the Moon as a living realm, as a place that could sustain human life for a period of time and as a source of potentially valuable natural resources. In such ways, the BIS space-ship represented a synthesis of astronautics and broader speculative understandings of outer space, considering not just the means of transportation in space, but also pondering some of the mysteries of the cosmos itself.

In the July 1939 edition of the *BIS Journal*, the last before the outbreak of the Second World War effectively suspended the Society's activities, William F Temple announced that the BIS space-ship proposal had been sent, with covering letters, to an array of well-known scientists, journals, magazines and newspapers. Temple reported that subsequent coverage by national newspapers had tended to emphasise the 'sensational aspects' of the plan, contrary to the Society's goal of prompting serious discussion of the proposal, but nonetheless, it had succeeded in raising the profile of the BIS, and reflected a growing public acceptance of the feasibility of spaceflight.[91] It was also reported that press cuttings were sent in to the Society from around the world, and it seems that the project had received substantial attention at this time, not entirely unfavourable. Less salubrious, however, was the reaction of the editors of the journal *Nature*, which rejected the proposal outright, noting that 'while the ratio of research results accomplished to speculative theorising is so low, little confidence can be placed in the deliberations of the British Interplanetary Society'.[92] This criticism echoes comments in an earlier edition of *Nature*, which dismissed Cleator's *Rockets Through Space* as 'a somewhat premature book on the possibilities of using rockets for interplanetary travel'.[93] Cleator and other members of the BIS had, by this time, become used to members of the scientific establishment rejecting concepts of spaceflight as fanciful, and this episode confirms that the Society's strategy of connecting with scientific audiences outside of the traditional echelons of power in science represented the best chance of successfully promoting the concept of interplanetary travel.

Conclusion: the spatiality of interplanetary science

Throughout its existence, the British Interplanetary Society has shaped technical, popular and organisational cultures of outer space, and this chapter has focussed on the Society's visions of space exploration by exploring its social, cultural and political geographies in the 1930s. These geographies varied in scale from the domestic to the international, and were instrumental in synthesising the science of outer space with the enduring dream of cosmic transcendence, engendering a particular type of socio-technical imaginary. Philip Cleator has emerged a key figure in British astronautics, his captivation with the idea of interplanetary travel motivating him to establish the BIS in 1933. Cleator and the Society's second President A M Low were charismatic individuals who drew productively from their connections in the scientific, journalistic and publishing worlds. Indeed, the Society had local, national and international connections with astronautics researchers and science-fiction communities, while its members met informally in the modest spaces in and around Liverpool and London that would accommodate their needs. Hindered by lack of finance and obstructive legislation, the Society eschewed practical experimentation in favour of international networking alongside 'big picture' theorising about space exploration and its implications, advancing several ideas that were to become central to the development of astronautics, some of which were exemplified in its ambitious 'BIS space-ship' project in 1939.

In focussing on the spaces of science associated with the BIS, two scales of engagement have been identified, each with particular implications for the ways in which astronautics was conceptualised in Britain. Firstly, the convivial spaces of science in which the Society was active enabled a collaborative interface between technical ideas and idealistic concepts of space exploration, shaping the BIS in ways that contrasted with the other principal spaceflight societies of the period in Germany and America. Second, the ways in which the BIS dealt with concepts of internationalism not only allowed a fruitful exchange of ideas, experimental data and contacts across international borders, but it also changed the ways in which astronautics was conceived, towards a science that was internationally minded in terms of the languages and philosophies it aspired to in this most idealistic period in the history of space exploration. The spatiality of interplanetary science, then, formed an important part of the cultures and politics of outer space in pre-war Britain, in which the desire to explore outer space was reconciled with the intractable mysteries of the cosmos.

Notes

1 Roger Lancelyn Green, *Into Other Worlds – Space-Flight in Fiction, from Lucian to Lewis* (London: Abelard-Schuman, 1958).
2 Frank Winter, *Prelude to the Space Age – the Rocket Societies: 1924–1940* (Washington DC: Smithsonian Institution Press, 1983).
3 Bob Parkinson, ed., *Interplanetary – a History of the British Interplanetary Society* (London: British Interplanetary Society, 2008).
4 David N Livingstone, *Putting Science in Its Place – Geographies of Scientific Knowledge* (Chicago: University of Chicago Press, 2003), p.179.
5 Simon Naylor, 'Introduction: Historical Geographies of Science – Places, Contexts, Cartographies', *The British Journal for the History of Science,* 38 (2005), 1–12; Christy Collis and Klaus Dodds, 'Assault on the Unknown: The Historical and Political Geographies of the International Geophysical Year (1957–8)', *Journal of Historical Geography,* 34 (2008), 555–573; Rory Mawhinney, 'Astronomical Fieldwork and the Spaces of Relativity: The Historical Geographies of the 1919 British Eclipse Expeditions to Príncipe and Brazil', *Historical Geography,* 46 (2018), 203–238; Diarmid Finnegan, 'Finding a Scientific Voice: Performing Science, Space and Speech in the 19th Century', *Transactions of the Institute of British Geographers,* 42 (2017), 192–205.
6 Rebekah Higgitt, *Recreating Newton: Newtonian Biography and the Making of Nineteenth-Century History of Science* (London: Pickering and Chatto, 2007); Jenny Uglow, *The Lunar Men – the Friends Who Made the Future, 1780–1810* (London: Faber and Faber, 2002); James Mussell, *Science, Time and Space in the Late Nineteenth-Century Periodical Press: Movable Types* (Aldershot: Ashgate Press, 2007).
7 Bruno Latour, *Science in Action: How to Follow Scientists and Engineers through Society* (Cambridge, Mass.: Harvard University Press, 1987); Michel Foucault, 'Of Other Spaces', *Diacritics,* 16 (1986), 22–27.
8 K Maria D Lane, in Oliver Dunnett and others, 'Geographies of Outer Space: Progress and New Opportunities', *Progress in Human Geography,* 43 (2019), 314–336 (p.324).
9 K Maria D Lane, *Geographies of Mars: Seeing and Knowing the Red Planet* (Chicago: University of Chicago Press, 2011).

10 K Maria D Lane, 'Astronomers at Altitude – Mountain Geography and the Cultivation of Scientific Legitimacy', in *High Places – Cultural Geographies of Mountains, Ice and Science*, ed. by Denis Cosgrove and Veronica Della Dora (London: I B Tauris, 2008), 126–144.

11 Denis Cosgrove, 'Contested Global Visions: One-World, Whole-Earth, and the Apollo Space Photographs', *Annals of the Association of American Geographers*, 84 (1994), 270–294.

12 Tobias Boes, 'Beyond Whole Earth: Planetary Mediation and the Anthropocene', *Environmental Humanities*, 5 (2014), 155–170.

13 Fraser MacDonald, 'Geopolitics and "the Vision Thing": Regarding Britain and America's First Nuclear Missile', *Transactions of the Institute of British Geographers*, 31 (2006), 53–71; Peter Redfield, *Space in the Tropics – from Convicts to Rockets in French Guiana* (Berkeley: University of California Press, 2000).

14 Fraser MacDonald, *Escape from Earth – a Secret History of the Space Rocket* (London: Profile, 2019).

15 Asif Siddiqi, *The Red Rockets' Glare – Spaceflight and the Soviet Imagination, 1857–1957* (Cambridge: Cambridge University Press, 2010).

16 Adrian Desmond, 'Redefining the X Axis: "Professionals", "Amateurs" and the Making of Mid-Victorian Biology – A Progress Report', *Journal of the History of Biology*, 34 (2001), 3–50 (p.4); James Secord, 'Knowledge in Transit', *Isis*, 95 (2004), 654–672.

17 Michael Neufeld, 'Weimar Culture and Futuristic Technology: The Rocketry and Spaceflight Fad in Germany, 1923–1933', *Technology and Culture*, 31 (1990), 725–752 (p.728).

18 Charlotte Sleigh, 'Science as Heterotopia: The British Interplanetary Society before WWII', in *Scientific Governance in Britain, 1914–79*, ed. by Don Legett and Charlotte Sleigh (Manchester: Manchester University Press, 2016), 217–233.

19 Alexander Geppert, 'Space Personae: Cosmopolitan Networks of Peripheral Knowledge, 1927–1957', *Journal of Modern European History*, 6 (2008), 262–286.

20 Nick Bingham, 'Socio-Technical', in *Cultural Geography – A Critical Dictionary of Key Concepts*, ed. by David Atkinson and others (London: I B Tauris, 2005), 200–212 (p.201); Sheila Jasanoff, 'Science, Technology, and the Imaginations of Modernity', in *Dreamscapes of Modernity: Sociotechnical Imaginaries and the Fabrication of Power*, ed. by Sheila Jasanoff and Sang-Hyun Kim (Chicago: Chicago University Press, 2015), 1–33.

21 Peter Bowler, *A History of the Future: Prophets of Progress from H G Wells to Isaac Asimov* (Cambridge: Cambridge University Press, 2017).

22 Winter, *Prelude to the Space Age*.

23 'What the Clubs Are Doing', *Practical Mechanics* 1 (1933), p.151.

24 Philip E Cleator, 'The Possibilities of Interplanetary Travel', *Chambers's Journal*, January 1933 (1933), 49–54 (p.54).

25 Philip E Cleator, *Rockets Through Space, or, the Dawn of Interplanetary Travel* (London: Allen + Unwin, 1936).

26 'British "Rocketeers" – Reaching the Moon – And Elsewhere', *Liverpool Echo*, 8th September 1933, p.10.

27 Leslie Johnson, *The British Interplanetary Society (1933 to 1945)*, p.5 [Unpublished memoir].

28 Winter, *Prelude to the Space Age* (p.148).

29 Charles Davy, 'Travelling Across Space – Rockets, Science and Enthusiasm', *Yorkshire Post*, 24th February 1936, n.p.; *New York Post*, 29th July 1934, n.p.

30 Philip E Cleator, *Letters from Baltimore – the Mencken–Cleator Correspondence* (London: Associated University Press, 1982).

31 David Kirby, *Lab Coats in Hollywood – Science, Scientists and Cinema* (Cambridge, MA: MIT Press, 2010).

32 Ursula Bloom, *He Lit the Lamp – a Biography of Professor A M Low* (London: Burke, 1958), p.6.

33 Lord Brabazon of Tara, 'Introduction', in *He Lit the Lamp – a Biography of Professor A M Low*, ed. by Ursula Bloom (London: Burke, 1958), 11–12 (p.11); Philip E Cleator, 'Obituary – Professor A M Low', *Journal of the British Interplanetary Society,* 51 (1956), 351.

34 Bloom, *He Lit the Lamp.*

35 Peter Bowler, 'Experts and Publishers: Writing Popular Science in Early-Twentieth Century Britain, Writing Popular History of Science Now', *British Journal for the History of Science,* 39 (2006), 159–187 (p.171).

36 Philip E Cleator, 'Terminal Testimony', *Journal of the British Interplanetary Society,* 39 (1986), 147–162 (p.151); Vauxhall, London, British Interplanetary Society Archive, Val Cleaver Box File, 1st February 1938.

37 Peter Bowler, 'Parallel Prophecies: Science Fiction and Futurology in the Twentieth Century', *Osiris,* 34 (2019), 121–138.

38 Geppert, 'Space Personae' (p.265).

39 Harry E Ross, 'Gone with the Efflux', *Journal of the British Interplanetary Society,* 9 (1950), 93–101 (p.97).

40 Anne Secord, 'Science in the Pub: Artisan Botanists in Early Nineteenth-Century Lancashire', *History of Science,* 32 (1994), 269–314; Larry Stewart, 'Other centres of calculation, or, where the Royal Society didn't count: Commerce, coffee-houses and natural philosophy in early modern London', *The British Journal for the History of Science* 32 (1999), 133–153.

41 Secord, 'Science in the Pub' (p.272).

42 Finnegan, 'Finding a scientific voice' (p.192); James Secord, 'How Scientific Conversation Became Shop Talk', *Transactions of the Royal Historical Society,* 17 (2007), 129–156.

43 Cleator's home was destroyed by a bomb in 1941 as the Luftwaffe targeted the Merseyside industrial and shipping area during the Second World War. See Cleator, 'Terminal Testimony' (p.149).

44 Cleator, *Letters from Baltimore* (p.29).

45 Johnson, *The British Interplanetary Society* (p.5).

46 Johnson, *The British Interplanetary Society* (p.6).

47 Philip E Cleator, 'Matters of No Moment', *Journal of the British Interplanetary Society,* 9 (1950), 49–53 (p.51).

48 Johnson, *The British Interplanetary Society* (p.23).

49 Johnson, *The British Interplanetary Society* (p.38).

50 *Daily Sketch*, 21st February 1936, n.p. [BIS 'Press Cuttings' book, BIS Archive, Vauxhall, London]; Johnson, *The British Interplanetary Society* (p.39).

51 Arthur C Clarke, *Astounding Days: A Science Fictional Autobiography* (London: Victor Gollancz Ltd., 1989), p.148.

52 Ross, 'Gone with the Efflux', (p.95–6).

53 Val Cleaver Box File, 1st February 1938; Vauxhall, London, British Interplanetary Society Archive, Val Cleaver Box File, 17th July 1938.

54 'Historical Geographies of Internationalism', *Political Geography* 49 (2015) [special issue]; 'Science and Geopolitics: The International Geophysical Year, 1957–8', *Journal of Historical Geography* 34 (2008) [special issue].

55 Jake Hodder, Stephen Legg, and Michael Heffernan, 'Introduction: Historical Geographies of Internationalism, 1900–1950', *Political Geography,* 49 (2015), 1–6. (p.4, p.3).

56 Martin Mahony, 'Historical Geographies of the Future: Airships and the Making of Imperial Atmospheres', *Annals of the Association of American Geographers,* 109 (2019), 1279–1299 (p.1285).

57 Arthur C Clarke, 'The Challenge of the Space Ship', *Journal of the British Interplanetary Society,* 6 (1946), 66–81 (p.76).

58 Cleator, 'Terminal Testimony' (p.147).

59 Oliver Sykes and others, 'A *City Profile* of Liverpool', *Cities,* 35 (2013), 299–318.

60 Everett Franklin Bleiler, *The Gernsback Years – A Complete Coverage of the Genre Magazines* Amazing, Astounding, Wonder *and others from 1926 through 1936* (Kent, Ohio: Kent State University Press, 1998).

61 Clarke, *Astounding Days.*

62 Mike Ashley, *The Time Machines – the Story of the Science-Fiction Pulp Magazines from the Beginning to 1950* (Liverpool: Liverpool University Press, 2000).

63 Johnson, *The British Interplanetary Society* (p.1).

64 'International Interplanetary News', *Journal of the British Interplanetary Society,* 2 (1934), 16–18.

65 Otto Steinitz, 'A Mistake of the Antagonists of Space Travel', *Journal of the British Interplanetary Society,* 1 (1934), 35–36.

66 Cleator, *Rockets Through Space* (p.146).

67 Cleator, *Rockets Through Space* (p.174).

68 Siddiqi, *The Red Rockets' Glare*; Neufeld, 'Weimar Culture and Futuristic Technology'.

69 Philip E Cleator, 'Editorial', *Journal of the British Interplanetary Society,* 1 (1934), 14 (p.14).

70 Cleator, 'Editorial' (p.16).

71 Cleator, 'Editorial' (p.14).

72 Johnson, *The British Interplanetary Society* (p.36); *Journal of the British Interplanetary Society,* 4 (1937), p.13; 'Notes and News', *Journal of the British Interplanetary Society,* 3 (1936), p.23.

73 Roberto Garvía, *Esperanto and its Rivals – The Struggle for an International Language* (Philadelphia: University of Pennsylvania Press, 2015), p.2.

74 Hodder, Legg and Heffernan, 'Introduction: Historical geographies of internationalism' (p.1).

75 Winter, *Prelude to the Space Age* (p.13).

76 Archibald Montgomery Low, 'Pioneering', *Journal of the British Interplanetary Society,* 4 (1939), p.4.

77 Low, 'Pioneering' (p.4).

78 William F Temple, 'Editorial', *Journal of the British Interplanetary Society,* 5 (1939), p.2.

79 Olaf Stapledon, in Temple 'Editorial' (p.2).

80 Ralph A Smith, 'Policy of the BIS', *Journal of the British Interplanetary Society,* 4 (1939), p.7.

81 Temple 'Editorial' (p.3).

82 Sleigh, 'Science as heterotopia'.

83 Harry E Ross, 'The Bis Space-Ship', *Journal of the British Interplanetary Society,* 5 (1939), 4–9 (p.4).

84 Smith, 'Policy of the BIS' (p.8).

85 Maurice K Hanson, 'The Payload on the Lunar Trip', *Journal of the British Interplanetary Society,* 5 (1939), p.13.

86 Hanson, 'The Payload on the Lunar Trip' (p.15).

87 Ralph A Smith, 'The BIS Coelostat', *Journal of the British Interplanetary Society,* 5 (1939), 22–27.

88 Hanson, 'The Payload on the Lunar Trip' (p.12).

89 In 2019, curators at the UK's National Space Centre commissioned a full-scale model of the 'BIS Space-Suit', which was designed by R A Smith and H E Ross in 1949. It is currently on display at the Centre's Rocket Tower exhibition area: Dan Kendall, 'The BIS Lunar Spacesuit', *National Space Centre* (2020) https://spacecentre.co.uk/blog-post/the-bis-lunar-spacesuit/ [7th August 2020].

90 Hanson, 'The Payload on the Lunar Trip' (p.16).

91 Temple, 'Editorial' (p.1).

92 'Interplanetary Travel', *Nature*, 143 (1939), p.635.

93 'Interplanetary Travel', *Nature*, 137 (1936), p.442.

4 Outer space and popular culture in post-war Britain

The post-war period was a time in which the concept of spaceflight became a tangible reality, while acting as a further spur to the imaginative representation of outer space. As scientists, military leaders and governments began to realise in the late 1940s that the technology of the rocket held the potential both for 'transcendence and mass destruction', a multitude of imaginative possibilities were also being expressed in diverse cultures of outer space, influenced by wartime developments in science and technology, earlier speculative fiction, as well as a range of cultural traditions and experiences.[1] Whereas such expressions had in common a blending of imaginative, popular and scientific cultures, this chapter argues that post-war understandings of outer space in Britain had a distinctive spatiality to them, both in terms of the grounded geographies of national, regional and local identity, as well as the more specific spatial contingencies of their production and reception. Focussing on similar themes in the context of the United States, the literary scholar De Witt Douglas Kilgore has identified a cultural movement of 'Astrofuturism', that took a liberal outlook on the possibilities of outer space technology, was couched distinctively in a Euro-American worldview, and transported a benevolent historiography of Western imperialism into the Space Age.[2] Kilgore charts the development of this discourse across the twentieth century, but explains how it came of age in the post-war period, when the influence of science-fictional imaginations on the emerging technocracy of the Apollo era was seen clearly through the likes of Robert Heinlein, Arthur C Clarke and Wernher von Braun. By contrast, this chapter, with its focus on the UK, argues that outer space can be understood not just as a space into which ideas of Western liberalism can be projected, but as a space that interfaces with a more particular set of social and cultural circumstances, including the distinctive British experiences of the Second World War, the impacts of decolonisation, the growth of domestic and popular cultures of scientific entertainment, and the ongoing influence of evolutionism and the sublime in science, culture and politics.

Contrary to suggestions that the post-war years in Britain were characterised mainly by austerity and cultural stagnation, this chapter identifies a multitude of engagements with outer space that take place from a range of ideological and geographical perspectives, from active interplanetarism to mythological readings of the cosmos. These include the expansionary, exploratory and sometimes utopian visions of outer space, represented in this chapter by the early space novels of Arthur C Clarke and the popularisation of astronomy by Patrick Moore; as well as more ambivalent cultures of outer space, including dystopian fictions and mystical encounters with the alien other, exemplified here by the *Quatermass* BBC television series, and the popular phenomenon of the Unidentified Flying Object. While these examples were chosen to demonstrate a broad thematic range of cultural engagements, other icons of outer-space culture in post-war Britain, such as Dan Dare and Doctor Who, which have been well-researched by scholars in cultural history and other areas, will form part of the context for interpreting the case studies that this chapter primarily focusses on. Over and above a more general appreciation of national, regional and local identity, these examples take us through a series of distinct spatial registers, emphasising geography's significance in the understanding of outer space. These include the earthly spaces of storytelling in the *Quatermass* television series, the aerial spaces of observation in the post-war UFO phenomenon, the domestic spaces of popular astronomy promoted by Patrick Moore, and the anticipatory spaces of interplanetary exploration in Arthur C Clarke's early spaceflight novels. Through these examples, understandings of outer space emerged in widespread and varied popular cultures for the first time in the UK, in ways that were distinctive and uniquely related to British science and society.

Outer space, myth and modernity in *Quatermass*

From 1953 to 1958, a popular new science-fiction serial, *Quatermass*, was broadcast on BBC television, an inventive and ground-breaking programme that established television's capacity to produce innovative outer-space imaginaries. Following the investigations of the eponymous professor, three of the classic tropes of science-fiction provided the narrative basis of the original *Quatermass* series: The exploration of space in *The Quatermass Experiment (1953)*, an alien invasion in *Quatermass II (1955)*, and the implication of extraterrestrials in human evolution, in *Quatermass and the Pit (1958)*. Interpreting the production context of *Quatermass*, alongside its reception by viewers and critics, helps to explain the ways in which narratives of myth and superstition, as well as icons of modernity, were implicated in outer-space imaginaries, while also enabling specific insights into the social and material environs of post-war Britain.

While not the first example of science-fiction in British television's pre-commercial era (earlier programmes included the children's serial *Stranger from Space* that ran from 1951 to 1953), *Quatermass* was the first to be aimed at a mainstream adult audience, and with its Saturday evening broadcast slot, in this and many other ways it was a forerunner to the 1960s' *Doctor Who*

Figure 4.1 Nigel Kneale on the set of Quatermass II, 1955

Source Credit: BBC Photo Library

and a long tradition of British science-fiction television.[3] Written by Nigel Kneale (1922–2006), who went on to become one of British television's most acclaimed writers (Fig.4.1), and directed by Rudolph Cartier, Quatermass is said to have 'revolutionised British television', and helped to elevate science fiction to a serious and substantial entertainment genre.[4] Watched by audiences that rose from an initial three million to eleven million by the end of the final series, these programmes were phenomenally popular, inspiring film adaptations by Hammer studios in the 1960s, as well as further television and radio re-makes, and they are now regarded as cult classics. Over and above this popularity and critical success, the *Quatermass* series is significant because it exemplifies the ways in which imaginative geographies of the sublime that were introduced in earlier fictive representations of outer space (see Chapter 2) continued to hold a strong influence in the post-war period, at times helping to subvert popular utopian imaginations of outer-space futures. In exploring these themes, *Quatermass* also demonstrates how the geographies of outer space can become wrapped up in the grounded social and (geo)political issues of the time, including post-war reconstruction, immigration, and the changing role of science in society.

Characteristic of Nigel Kneale's writing throughout his career was the interplay between the promise of modernity and the mysteries of the deep past, as he mixed elements of science-fiction with aspects of folk horror. This hybridity is evident throughout the narrative of *The Quatermass Experiment*, which opens with scenes evoking the spectacle of spaceflight and ends with the horror of the monstrous alien other. The first episode sees the 'British Experimental Rocket Group', led by Professor Quatermass, launch a manned rocket into space. Mimicking the ambitions of the British Interplanetary Society, the launch takes place from 'Tarooma Base' in Australia, a pastiche of the actual rocket testing range that was at this time active in Woomera, South Australia (see Chapter 5). The opening scenes feature stock footage of a V–2 rocket launch, seen from both ground-level and an on-board camera, showing the receding Earth and the emerging blackness of space. By the early 1950s, American developments in rocketry, based on augmented versions of the V–2, had resulted in the release of such imagery to media organisations, fuelling speculation that the advent of spaceflight was imminent and cementing the space-rocket as an icon of twentieth-century modernity.[5] However, Kneale upended this expectation by having his rocket crash-land in a London suburb, echoing the real devastation caused by the V–2 less than a decade earlier. It transpires that the three astronauts on-board have been exposed to an alien life-form in space and have had their consciousness merged into one figure, superficially the astronaut Victor Carroon. After emerging from the wreck, Carroon transforms physically, becoming a hybrid plant-human-fungoid creature that terrorises the streets of London, and finally comes to rest in the high alcoves of Westminster Abbey to complete its metamorphosis.

Part of Kneale's inspiration for the more subversive and darker aspects of his writing can be traced to his connection to the Isle of Man and associated folk cultures of storytelling. Although he was born in the town of Barrow-in-Furness in Cumbria, Kneale's parents were originally from the Isle of Man, and they returned there when he was six years old, with Nigel's father taking up the editorship of the island's main newspaper. Kneale's biographer Andy Murray notes how he 'grew up steeped in ancient Manx superstition', the Irish Sea island being culturally and politically separated from the United Kingdom, while also being home to various Celtic and Norse traditions.[6] Here, fearful stories of creatures such as the shape-shifting *Buggane* and the nocturnal *Phynodderee* are still recited as part of Manx folklore.[7] Such legends found their way into Kneale's consciousness:

> It was about showing due respect for things that were not entirely to be understood, namely this score of wild superstition [*sic*] creatures who had grown out of the island's soil, practically, up in the mountains.[8]

Traits of the organic and the monstrous can be seen in the alien hybrid creature that gradually emerges throughout *The Quatermass Experiment*, in what Kneale later called 'the ultimate and unclassifiable monstrosity'.[9] Kneale knowingly set the final episode in Westminster Abbey, which just weeks before its broadcast had been the setting of Queen Elizabeth II's coronation, the first mass-participation event in the history of British television. While not allowed to film in the Abbey itself, Kneale shot scenes against blown-up publicity photos of its interior, through which he ingeniously inserted a home-made monster made out of a pair of old leather gloves and some foliage, an effect that he later affirmed as 'extremely sinister'.[10] With such limited support in special effects production, Kneale had to rely on his audience's capacity to imagine, and he maintained that the public memory of the coronation 'lingered' in viewers' minds, helping to create the illusion of an alien life-form in Westminster Abbey.[11] Here, it was the benign modernity associated with the 'TV coronation' that Kneale was able to subvert through his audacious insertion of the unknowable alien monstrosity in a place that was so recently re-affirmed at the heart of British public life. Concluding the narrative of *The Quatermass Experiment*, Professor Quatermass is able to commune with the human vestiges of the creature, persuading it to destroy itself rather than allow its spores to spread across London. This anti-heroic ending echoes H G Wells, whose Martians finally succumb to bacteria and wither away in *The War of the Worlds*, but is perhaps more ambiguous in its final resolution. Indeed, for critics Rolinson and Cooper, the narrative owes much to Mary Shelley's *Frankenstein*, the archetypal Gothic horror story portraying the dark repercussions of modern science.[12]

In such ways, Kneale's teleplay presents space exploration in ways that might not have been expected by audiences. Whereas much science-fiction of the early-post-war period involved the heroic conquest of space, *The Quatermass Experiment* starts with a catastrophic failure, incorporates themes of body-horror, and ends in desperate self-destruction. Part of the reasoning behind this was to provide a counterpoint to American science-fiction narratives, which are parodied by Kneale with a scene inside a cinema showing the apocryphal title 'Planet of the Dragons'. More fundamentally though, Kneale sought to exploit ideas that contained contradictions, with *The Quatermass Experiment* pivoting around the notion that 'we go into space, making progress, but bring something back which takes us backwards'.[13] This theme is continued in *Quatermass II*, in which a malevolent alien life-form infiltrates the minds of government officials who are investigating falling meteors in the British countryside. Certainly, the idea that outer space could be threatening, unknowable and horrific ran counter to much of the optimism found in post-war narratives of space exploration (see, for example, the Clarke case study in this chapter) and brings a different dimension to understandings of outer space in this period.

While this melancholic tone was likely influenced by the folk tales of Kneale's Manx heritage, as well as broader Gothic and Romantic traditions, critics have suggested that *Quatermass* also reflected the traumatic public experience of the Second World War, with much of Britain still bearing its physical and psychological scars.[14] Kneale volunteered for military service during the war, but was granted a medical exemption due to a form of photophobia, his skin reacting badly to sunlight exposure.[15] This was a factor that led Kneale towards writing rather than directing or acting in the post-war television industry and perhaps also contributed to themes of bodily metamorphosis in the *Quatermass* narratives. While Kneale had no direct experience of the war on the continent, he visited London in 1943 to sign a writing contract and returned permanently in 1946. There he witnessed some of the carnage caused by 'The Blitz' of 1944–1945, including thousands of destroyed buildings and the ongoing danger of unexploded bombs, finding 'a city trapped in the post-war stasis'.[16] Aspects of this postwar landscape can be identified at key points in the *Quatermass* series, from the suburban crash-site at the beginning of *The Quatermass Experiment*, to the crater at the epicentre of *Quatermass and the Pit*. The former has been described in terms of its 'wartime symbolism, with locals rallying around an old woman who stoically accepts the demolition of her house', recalling aspects of the mythologised 'Blitz spirit'.[17] Kneale latched onto the more general concept that 'people were frightened of what might drop from the skies', in dealing with feelings of malaise in post-war society, whether associated with the social memory of The Blitz or the anticipation of nuclear bombardment that came with the onset of the Cold War.[18]

After destruction comes re-building, and in *Quatermass and the Pit*, the majority of the action takes place at an excavation site in London's Knightsbridge, in scenes that reflected ongoing activity in the capital's post-war urban landscape. Here, Professor Quatermass discovers a host of preserved insectoid Martian creatures within a buried space vessel in 'the pit', which is initially declared as the location of an unexploded bomb. London becomes again the focus of a narrative pivot between lived reality and the altered state of the science-fictional imagination, and whereas for H G Wells half a century earlier, London's scientific urban landscape inspired a dystopian take on the evolutionary future of life on Earth, for Kneale, the wartime destruction of the capital city provided a portal into a disturbing evolutionary past. In this way, some of the tangible dangers of life in postwar Britain were re-conceptualised as fear of the alien unknown, and the even more disturbing realisation that Martian creatures had intervened in humanity's deep past and psychic evolution. Such themes were picked up on by one reviewer at the time, who noted the 'unearthly echoes of horrors to come' in the first episode of the series, while concepts of the evolutionary depths of humanity bring to mind the sublime realisation of deep time that was acknowledged in the earlier narratives of Olaf Stapledon.[19]

Kneale later explained the storyline to *Quatermass and the Pit* as a metaphor for racism in society, shocked as he was by the 1958 race riots in parts of London and Nottingham that followed the Windrush-era immigration of workers to Britain from its former colonies.[20] For Kneale, exposing the grand narratives of human evolution, and suggesting its possible alien origins, brought into focus the futility of racism in contemporary society, a framework that stands in contrast to the racial politics of American Astrofuturism.[21] One report of public reaction to this theme in *Quatermass and the Pit* came from Birmingham, where leading members of 'the West Indian community' criticised the programme for its portrayal of a fictional news bulletin that stated 'race riots are continuing in Birmingham', which was said to have created a false impression of racial unrest in the city.[22] Whereas the BBC rejected the complaint on the grounds that it was a completely imaginary news bulletin, its reception as a relevant indicator of societal issues emphasises the grounded reality of Kneale's fantasy narratives. This serves as an example of how, although the teleplays of *Quatermass* incorporated social realism, their fantasy elements helped to conceptualise the ways in which human trauma was being processed at a deeper level, not only the ordeal of war, but also the scourge of violence and hatred in society more generally. Writer Mark Chadbourn interprets this via Jacques Derrida's concept of 'Hauntology', suggesting that Kneale 'mapped Britain's unconscious at a time in our history when all sorts of lines were blurring', including divisions between the colonial and postcolonial, the deep past and an ambivalent technological future.[23]

Here, and throughout the *Quatermass* series, it is demonstrated that themes of modernity in science-fiction narratives were not incompatible with more complex expositions of myth, legend and superstition, with earth and organic matter given prominence as much as shiny metal spaceships. The impact of *Quatermass* in British post-war popular culture is evidenced by the mass viewership it achieved, as well as its continuous re-appraisal in television, film, radio and cultural criticism.[24] Part of the popular appeal of *Quatermass* also undoubtedly was connected to Kneale's humanistic approach to writing, focussing on the behaviour of ordinary people in extraordinary situations, and, again echoing Wells, following the mantra that 'the strangest things ought to happen in the most ordinary of places'.[25] This ethos helped to produce an iconic example of outer-space fiction that dealt uniquely with the tangible and psychological ambiguities of outer space, echoing earlier conceptions of the sublime in outer-space imaginaries. With its narrative themes ultimately aligned more to the mythological than the modern, *Quatermass* subverts not only the predominant Astrofuturist cultures of the time, but also some of the foundations of British society itself.

Interpreting the unknown: the UFO phenomenon in Britain

A central dichotomy of popular cultures of outer space, whether in television, literature or folklore, has been an acknowledgement that space can signify the limits of human understanding, while also acting as a source of transcendence from Earthly bounds. Encapsulating this joint appeal, a significant feature of post-war cultures of outer space in Britain has been the phenomenon of the UFO, the Unidentified Flying Object that is, in the minds of some, assumed to be of intelligent extra-terrestrial origin.

While the appellation 'UFO', coined by a US Air Force Captain in 1950, tacitly acknowledges the unknown, other names that have been used to denote such phenomena, including 'flying saucers' and 'ghost rockets', provide satirical or mystic suggestions to the objects' provenance. A range of amateur organisations have sought to ascertain the veracity of UFOs, thereby attempting to explain their mysteries. Such accounts have ascribed UFO sightings to astronomical and meteorological phenomena, various types of secret or military aircraft, hoaxes or even hallucinations, tending to conclude that the vast majority of UFO sightings have rational explanations. However, the few remaining unexplained sightings have continued to fuel a speculative 'extra-terrestrial hypothesis', perhaps influenced by science-fiction narratives or actual achievements in spaceflight. One commentator of the period gave a sense of the extent of the phenomenon, noting that 'between 1947 and 1952 some 3,000 flying saucers were reported to have been seen in various parts of the world'.[26] Evidence from surveys has since indicated the ongoing prevalence of UFO cultures, with large numbers of people having reported UFO sightings or believing in the past visitation of extra-terrestrials.[27] Sometimes involving 'contactee' or even alien abduction narratives, UFO cultures bring to bear a complex range of physical, technological, cultural, political and psychological factors, with implications for the ways in which people understand space exploration, life on other worlds and the possibility of extra-terrestrial intelligence.

Regardless of their ultimate veracity, it is important to acknowledge the UFO's role as 'an integral part of the history of spaceflight', with connections between UFOs, spaceflight promoters and popular culture propagating widely in the mid-twentieth century.[28] This reflects the context in which the UFO phenomenon emerged, following the popularisation of space exploration in science-fiction, coinciding with a post-war upsurge of public interest in spaceflight, and preceding the onset of the Space Age itself. A broader range of cultural and political factors have also been taken into consideration in UFO narratives, such as the accelerating decolonisation from European empires in the post-war period, the increasing secularisation of Western societies, and the onset of Cold War geopolitical tensions.[29] Acknowledging these contexts, researchers in the social sciences have tended to take UFO cultures at face value, being interested in the subjects of the sightings and the controversies they initiated, rather than assigning the objects any particular cause or explanation.[30] The psychoanalyst Carl Jung, for example, considered UFOs as objects that provoke 'conscious and unconscious fantasies' in their beholders, relating to a range of archetypal images in Western culture.[31] It has also been suggested that the persistence of paranormal beliefs in Western society signifies a 'haunted culture' that 'disturb[s] the ordered rationalism that comforts the late-modern mind', as similarly outlined above in relation to the *Quatermass* narratives.[32] In Britain, a leading researcher of UFO cultures has recognised 'Ufology' as part of folklore, looking to assign cultural or social meanings to the substantial archive of UFO reports that have been collected by public and private organisations since the Second World War.[33] The question therefore remains, what can UFO sightings tell us about British cultures of outer space, and their geographical meanings, in the post-war period?

In the first incidence of a UK government response to UFO sightings, the Ministry of Defence set up the 'Flying Saucer Working Party' in October 1950. This committee was established at the suggestion of Sir Henry Tizard, who was then the chief scientific adviser to the Royal Air Force, and chaired by G L Turney, the head of scientific intelligence at the Admiralty.[34] The Working Party's five-page report, released to the public only as recently as 2002, acknowledged media accounts of UFOs in northern Europe and America in the late 1940s, as well as two US government investigations into UFOs, 'Project Sign' and 'Project Grudge'. These occurrences have been noted elsewhere, with Sweden and Norway in particular becoming 'the theatre of nearly a thousand sightings of strange phenomena described as rockets of unidentified origin' in 1946, amid 'fierce debates about secret weapons and the fear of a third world war'.[35] A spate of UFO sightings in America in 1947 started with the account of Kenneth Arnold, who was flying his private aeroplane in Washington State when he perceived nine 'saucer like' aircraft flying in formation at great speed, his story being reported widely in the media and ultimately leading to the US Government investigations.[36] The interest of military intelligence agencies in UFOs has been linked to 'the possible defence threat posed by "unidentified" aircraft or missiles' in the context of the Cold War, and the MoD committee's acknowledgement of these events indicates the international scale of UFO cultures and their influence among Western states, not only in popular culture but also in government investigations.[37] Geopolitical considerations, therefore, played an important part in official understandings of UFO phenomena, whether through sharing of information between allied nations or in suspicion of international adversaries.

Having noted such prior incidents, The Flying Saucer Working Party's report focussed on an outbreak of UFO sightings in Britain during the summer and autumn of 1950. This coincided with growing public interest in UFOs, with two Sunday newspapers serialising two of the first books on UFOs around this time.[38] A number of sightings were investigated by the committee, including one account from a fireman in Derby, the object of which was determined to be a meteor, and three reports from Royal Air Force officers in southern England. Of the latter, there was one from the pilot of a Gloster Meteor flying above the city of Portsmouth, whose sighting appeared to have been confirmed by radar, and two from the ground at RAF Farnborough in Hampshire. One of these sightings was reported as follows:

> F/Lt Hubbard, who alone was wearing sun-glasses, states that he saw, almost directly overhead, at first sighting, an object which he describes

as a flat disc, light pearl grey in colour, about 50 feet in diameter at an estimated height of 5,000 feet. He stated that he kept it under observation for 30 seconds, during which period it travelled at a speed estimated at 800–1,000 mph, on a heading of 100⁰, executing a series of s-turns, oscillating so that light reflection came from different segments as it moved.[39]

Incidents such as these were attributed by the committee to aircraft, weather balloons, radar interference or other rational causes, and the report concluded that no further official investigations into UFOs should be carried out.

What is notable about the Flying Saucer Working Party's report is the way in which its conclusions were framed in the context of reliable forms of aerial observation. The committee affirmed the observers' integrity in each case, noting that even 'experienced observers' could be mistaken by optical illusions.[40] While not doubting the impression of F/Lt Hubbard's UFO sighting in the eyes of the observer, the report claimed that the absence of corroborating accounts 'over a populous and air-minded district like Farnborough' made it impossible to verify the pilot's observations.[41] The town of Farnborough in Hampshire, known for its annual air show since 1948 and as the location of the Royal Aircraft Establishment since 1904, was, in the minds of the committee, productive of the kind of citizen that would be particularly conscious of aerial space. This sense of 'airmindedness' in mid-twentieth-century Britain encapsulated aviation's potential to transform society, while also foreshadowing its 'darker connotations', including the threat of aerial bombardment.[42] Similarly, popular culture icons such as Dan Dare exemplified in this period the heroic reverence of the airman in post-war British culture, an archetype that in this character completed its logical extension to the realm of outer space.[43] This was also a time in which the aircraft industry was growing rapidly, as the Korean War, amid broader Cold War anxieties, helped fuel a massive re-armament programme in the UK.[44] Here the spectre of wartime bombing, both past and anticipated, alongside reciprocal associations of heroic air prowess, fed into the discourse that surrounded UFO sightings in post-war Britain, an appreciation that was assigned a particular significance in the case of the UFO sightings above Farnborough in southern England.

UFO sightings can also be understood as cultural signifiers for a wider register of interests, including spaceflight, extra-terrestrial intelligence and the mysteries of outer space. An ongoing fascination with UFOs was manifested in the establishment of the British UFO Research Association (BUFORA) in 1964, which remains one of the longest-running UFO research groups in the world. By the late 1960s, BUFORA's membership was said to have been approaching 750 individuals, with eighteen separate UFO societies conducting their own programmes of observation, reporting and research around the UK.[45] BUFORA's primary aim was to 'encourage and promote unbiased scientific investigation and research' into

UFOs, and its activities in the 1960s included monthly talks at Kensington Central Library in London, the publication of journals and bulletins, and the maintenance of a reference library for members.[46] In its activities and publications, BUFORA consolidated UFO culture in Britain and sought to legitimise UFO research as a rational science based on common standards of observation and evidence.

The quarterly *BUFORA Journal* reported sightings from around the UK and internationally, with an evaluation officer suggesting possible explanations for the causes of the phenomena. For example, an 'unusual flying object' with a 'dome-like protuberance' spotted in Epping in December 1963 was explained as an aircraft flying with its wing 'edge-on' to the observer.[47] However, in Macclesfield in August 1964, sightings of a 'silver coloured object', 'behaving erratically' with the shape of a 'shallow round dome', were designated as 'genuine UFO phenomena' by BUFORA's evaluation officer.[48] The same issue of the *BUFORA Journal* recounted sightings in Bedfordshire, Northumberland, Sussex, Nottingham, London and Cheshire, demonstrating a genuine popular fascination with the concept of UFOs. These reports also indicated BUFORA's willingness to accredit UFOs with possible extra-terrestrial provenance, posing an epistemic challenge to their strict adherence to observational accuracy and objectivity.

During this period, the *BUFORA Journal* also published letters, book reviews and short editorials on various space-related themes, including extra-terrestrial intelligence, spaceflight and astronomical observation, becoming a home for broader speculative discussions about outer space. One article outlined the possibilities of contact with extra-terrestrial intelligence and its implications for human spaceflight, considering the multiplicity of galaxies and planetary systems, while outlining an ethical protocol for contact with inhabited planets.[49] Observational astronomy also featured, with readers encouraged to familiarise themselves with the constellations of the night skies in order to 'minimise the chance of believing a known star or phenomenon to be a UFO'.[50] In this way various aspects of space science were discussed and disseminated in the context of UFO observation, comprising a sub-culture of outer space enthusiasm that was overlooked or dismissed by officials, professional science bodies and media commentators. In doing this, the *BUFORA Journal* also sought to prepare those interested in UFO culture for the prospect of transcendence, whether this would consist of catching sight of a genuine UFO in the night sky, or initiating discourse with prospective extra-terrestrial beings.

Interspersed among space-related articles and UFO reports in the *BUFORA Journal* were editorials that skirted a line between promoting UFO observation and defending its scientific validity. The journal's editors were convinced sincerely by evidence for the existence of 'genuine' UFOs, pointing towards the ten percent of sightings that could only be explained by the brazen simplicity of an extra-terrestrial hypothesis. At the same time, BUFORA was aware that UFO research attracted individuals described as 'emotionally unstable persons, cultists and space-struck juveniles', as it exhorted its members to maintain high standards of scientific accuracy in their observational practise.[51] As such, BUFORA represented 'mainstream' UFO culture in Britain, accepting the concept of extra-terrestrial provenance while rejecting some of the more bizarre forms of paranormal activity, perhaps even representing a 'haunted culture', trapped somewhere between rationalism and mysticism.[52]

UFO culture in post-war Britain came to exemplify various understandings of outer space, including a new awareness of aerial space as an arena of vigilance and potential conflict; an acknowledgement that modern, rational science could not answer all the questions posed by the mysteries of outer space; as well as a popular synthesis of space science involving astronomy, astronautics and life in the Universe. Indeed, while UFO culture helped to stimulate popular enthusiasm for space science, at the same time, it created a problem for the 'incipient space experts' of the post-war period, who were at this time attempting to make outer space 'socially respectable' in anticipation of a forthcoming Space Age.[53] The next sections of this chapter turn to two such figures, Patrick Moore and Arthur C Clarke, to examine their own distinctive cultural engagements at the interface between outer space, science and society in this period.

Domesticating outer space: Patrick Moore, Spaceflight and the Sky at Night

Part of the emerging popular culture of outer space in post-war Britain was a particular type of discourse that aimed to domesticate space science, making it familiar and practical while eschewing some of the more esoteric aspects of outer space culture. This discourse distanced itself actively from notions such as extra-terrestrial visitations to Earth and the impalpable horrors of deep space, seeking to make outer space knowable and tangible, within reach of human endeavour. A key figure was the astronomer, writer and television presenter Patrick Moore (1923–2012), whose work encapsulated a 'realistic and optimistic' approach to space science, and was made famous in Britain by the television programme *The Sky at Night*, which he presented from 1957 onwards.[54] Lacking any formal scientific training, Moore was an accomplished amateur astronomer, his maps of the Moon used by both the Russian and the American space programmes during the 1950s and 1960s.[55] Working across a range of media, including his non-fiction guidebooks, his editorship of popular science publications, and on television, Moore's dedication to amateur astronomy was demonstrated not only through his own observations, but in his exhortation of members of the British public to take up this activity themselves, and he was recognised for this work with a knighthood in 2001. As this section explains, the popularisation of space science in such ways implicated domestic space as the location of valid scientific activity, a theme picked up on in previous chapters but now given centre stage.

Moore was a long-standing member of the British Interplanetary Society, and when in 1955 the BIS decided to publish a popular magazine to help stimulate public interest in space exploration, to counterbalance its more technical *BIS Journal*, Moore was appointed as its first editor. The new publication, *Spaceflight*, followed in the tradition of popular science magazines in Britain, and became the only British magazine entirely devoted to space exploration when it was first released in 1956, on the cusp of the dawning era of space exploration. Although a popular publication, the tone of *Spaceflight* was one of scientific accuracy. Moore's opening editorial emphasised the need to counteract 'nonsense [...] misconceptions [...] and other space-borne crockery'.[56] Echoing the language of BIS founder Philip Cleator in the 1930s, Moore became an advocate for the spreading of 'correct information' and he appeared on BBC television in 1956, in a 'two-handed discussion' in which he argued against the existence of 'flying saucers'.[57] This factual approach was adopted in *Spaceflight*, and would be justified not simply for its own sake, but also to promote 'the layman's education', bringing about the possibility of armchair science at home for its readers.[58]

In line with this educational tone, the first issue of *Spaceflight* included a basic introduction to rocket technology, and reports from the latest American attempts to launch a satellite. Alongside these types of informative article were lighter pieces such as book reviews and cartoons, thus creating a 'mixed bag' that would offer a more general appeal.[59] Moore solicited and vetted the articles himself with the assistance of an editorial board, and he had overall control over *Spaceflight*'s output until he left the post in July 1959. Within the broader remit of popularisation, Moore's aims were;

> to do two things. First of all, to set up a basic information centre, to get all the information about space flight, and secondly, to set up satellite tracking stations around the world.[60]

Although the BIS never succeeded in setting up stations around the world, satellite-tracking was to become a prominent theme in *Spaceflight* once successful launches started to take place in the closing months of 1957. Before this point, readers were introduced to the concept of tracking objects in the sky through a regular series of instructive articles on astronomy under the title 'Sky Diary', foreshadowing Moore's later role in *The Sky at Night*. Moore noted that there was a large overlap between the astronautics and astronomy communities at this time, reflecting his belief that eventually 'rocketry and astronomy must merge into one science', and indicating the co-reliance of these two areas of space science, both of which could become the subject of amateur observation.[61]

Once artificial satellites did begin to appear in Earth orbit, readers were encouraged to observe the sky in order to track them. Even before the launch of *Sputnik* in October 1957, *Spaceflight* reported that the Yorkshire branch of the BIS was establishing a small observatory at Harrogate in anticipation of an American satellite launch as part of a world-wide amateur scheme for the International Geophysical Year (1957–8) called 'Moonwatch'.[62] Following the successful Soviet launch, press photographs were published in *Spaceflight* showing *Sputnik* leaving a trail across the night sky. These photographs would have suggested to the reader that observation of artificial satellites was possible from the ground, with equipment as simple as a camera. Indeed, the television aerial seen in one of the photographs is a recognisable element that would familiarise this perspective for the reader as a domestic gaze. Alongside the photographs, instructions were set out in the accompanying *Spaceflight* article, noting that the best chance of spotting the satellite would be when its passage over Britain came within one hour of sunrise or sunset, with the use of a pair of binoculars or a low-magnification telescope.[63] Apart from visual images representing *Sputnik,* the better-known moniker of this satellite was its iconic 'beep-beep' sound that was emitted by a radio transmitter, and there were efforts to track *Sputnik* by ham radio enthusiasts, both in the United States and in the Soviet Union.[64] However, these efforts are said to have been not nearly as effective as the visual observing programme, which by its nature was a simpler and more widely accessible method of participation.[65] This mode of public engagement, incorporating both spaceflight and astronomy, indicates an everyday, accessible means of amateur outer-space visualisation, in a manner which contrasts with the more passive 'Apollonian gaze' associated with picturing the Earth from space, that came to dominate outer-space imagery in the subsequent era of human spaceflight.[66]

Moore realised the potential of audience engagement to generate popular interest in outer space, and a key characteristic of his editorship of *Spaceflight* was encouraging readers to use information from the magazine in their own observational performances. This format was to be used to even greater effect in the medium through which a large proportion of the British public has come to know Patrick Moore, the BBC television programme *The Sky at Night*. Following his initial TV appearance in 1956, Moore contacted BBC producer Paul Johnstone with a suggestion for a new programme:

> The other scheme I did mean to suggest was a series devoted to practical astronomy, giving people ideas as to how they themselves can take up astronomy as a hobby and do observational work.[67]

It has been noted that the BBC from 1955 onwards sought to move away from traditional 'talks' delivered in the manner of a lecture, towards 'a mode that prominently featured scientists and technologists as television performers', and *The Sky at Night* took this development one stage further by enrolling the spirit of amateur astronomy and encouraging audience participation.[68]

The new series was broadcast into peoples' homes in a monthly fifteen-minute slot, usually after 10 pm on a week-night. This time-slot would correlate with the darkness of the night sky, while the programme's launch in 1957, like *Quatermass* four years earlier, took advantage of the upsurge in television ownership in Britain that followed the 1953 Coronation of Elizabeth II. Although the launch of *The Sky at Night* was not publicised heavily in the national press, the opening programme was estimated to have reached ten per cent of the UK adult population (or approximately 3.7 million viewers).[69] Once the format had been established, with Moore's personally-written scripts dictated live on air, it hardly changed during Moore's lifetime. The introductory music 'At The Castle Gate' by the late-Romantic Finnish composer Sibelius was followed by the title information, before cutting to the presenter, facing the viewer:

> What I want to do in these talks of mine is to tell you about some of the interesting things you can see in the night sky each month. Astronomy's not just a hobby for old men with white beards, as so many people think: everyone can take an interest in it – you don't need a vast telescope.[70]

A key part of this appeal was the promise that viewers could experience astronomy from their own homes and gardens, contrary to the assumption that astronomy was an activity that only took place in large professional observatories. Indeed, an audience survey carried out by the BBC noted that Moore's account of the Arend-Roland comet on the programme's first episode 'sent many viewers out-of-doors after the talk to do their own "spotting"'.[71] At the same time, Moore wanted to modernise the view of astronomy, away from 'white beard' associations towards a more widespread appeal that aimed 'to hit a middle course' between scientific lucidity and amateur interest, much akin to the tone of *Spaceflight*.[72] Moore himself was 34 years old when the first broadcast went out, an engaging figure, and he certainly didn't have a white beard.

Once the programme had been introduced, Moore used a variety of techniques to demonstrate and explain the phenomena that could be seen in the night sky (Fig.4.2). Studio graphics, or 'Wurmsers', were used to illustrate some of the key constellations, and direct viewers to other objects of interest:

> Saturn is certainly worth looking at […] remember, take a line from the Great Bear's tail, and you'll come to it.[73]

In this way, the Great Bear became known as 'a sort of sky signpost', as was the other recognisable constellation Orion.[74] A further characteristic of *The Sky at Night* was the range of experts who were invited on to the programme to explain various phenomena in an authoritative and informed manner. Early on, Moore suggested this idea to Johnstone:

> You may consider that it is worthwhile to get a biologist to show that the idea of human beings on Venus or Mars is quite untenable.[75]

Figure 4.2 Patrick Moore pointing out the position of Capella on *The Sky at Night*,
1960

Source Credit: BBC Photo Library

In a recurrent theme, Moore wanted to promote a realistic view of the Solar
System, and humanity's likely role in its exploration. Having invited the
American astronomer Harlow Shapley on to the programme in September
1958, the possibility of life on other planets was discussed amid some modest, domestic props including a bookshelf, a curtain and some informal
seats, which once again would have emphasised the suggestion that interplanetary science was something that could be engaged in from the comfort
of one's home. In this way, the use of domestic spaces of outer-space enthusiasm continued the tradition of amateur engagement that was found in the
activities of the pre-war BIS, which held meetings, demonstrations and talks
in a variety of home and public spaces (see Chapter 3). On the programme,
Shapley spoke of 'a high probability that there is abundant life scattered
about the universe', while Moore concluded the discussion by stating that
'where life can appear, life will appear'.[76]

It was generally down to Moore himself to invite guests on to the programme,
and the range of guests included BIS members such as Philip Cleator, Arthur
C Clarke and Val Cleaver.[77] Emphasising this connection, Clarke appeared in
1963 to talk about the work of the BIS in the 1930s, communication satellites,

and the possibilities for manned lunar bases.[78] Through this range of guests and subject-matter, along with the encouragement of audience participation, *The Sky at Night* promoted all aspects of space science, including human space travel, conditions on other planets and the tracking of artificial satellites. That audiences responded to this format is evident through the 'vast amounts of letters' that Moore received from viewers, while on one occasion 10,000 viewers wrote in to request 'star maps'.[79] It is also clear that what held all these elements together was the performance of Patrick Moore in front of the camera, and it quickly became apparent to producers of the programme that '*The Sky at Night* is really the Patrick Moore show'.[80]

Patrick Moore played a key role in the popularisation of space science in Britain during the early post-war period, through these two popular cultural outlets. Common themes included the encouragement of audience and reader participation in observing objects in space from domestic settings, particularly after the first satellites were launched, and the combination of astronomy and astronautics, with the space hardware of rockets and satellites being given roughly the same level of attention as the stars and planets in the night sky. While these activities undoubtedly popularised space science, they also assumed a certain middle-class privilege in their anticipated participants, who would have the time and space in which to engage with such emerging cultures of leisure in post-war Britain. While Moore, on the face of it, promoted factual and rational understandings of outer space through these media, he was also involved in a number of more whimsical projects, including the Irish alien B-movie *Them and the Thing* (1956) and UFO book *Flying Saucer from Mars* (1954). Researchers have claimed that the supposed author of this text, Cedric Allingham, was in fact Moore himself, playing out an elaborate hoax.[81] Curiously, when he appeared on the BBC debate programme *First Hand* in 1956, Moore cited the story by 'the late Cedric Allingham', as an example of easily-identifiable hoaxes that whipped up public interest in UFOs, as he maintained that there was no evidence to support the existence of extra-terrestrial UFOs.[82] It appears that, while Moore strongly advocated a factual understanding of the Universe, he also indulged in fantasy narratives of alien visitations and space adventures, either to prove a point about their falsifiability or perhaps to satisfy a yearning for transcendence himself. This postscript to Moore's engagement with outer-space cultures indicates how the lines between fantasy and reality, science and fiction, were not so easily drawn after all, a notion that can be addressed more directly through examining the works of one of the best-known British science-fiction writers of the twentieth century.

The language of interplanetarism in Arthur C Clarke's space trilogy

Known as one of the 'big three' science-fiction writers of the twentieth century, alongside Americans Isaac Asimov and Robert Heinlein, Arthur C Clarke made connections between the worlds of popular science, science fiction and space science, becoming a key figure in post-war British cultures of outer space, as well as the wider transatlantic discourse of Astrofuturism. Central to this interchange was Clarke's involvement with the British Interplanetary Society in London, where he acted as Chairman in 1946–1947 and 1951–1953, and at which time he formulated ideas that would be central to his long-term perspectives on spaceflight, technology and the future. In terms of his writing career, Clarke graduated from short stories in science-fiction 'pulp' magazines to novels published on both sides of the Atlantic by the 1960s. During this period of intensive writing, Clarke helped to develop a language of space exploration, repurposing the narrative tropes of the maritime adventure and other genres of fiction to promote a discourse of interplanetarism that had strong connections to the emerging science of outer space, arbitrated through the forums of the BIS. These themes had an implicit geography to them: the imaginative spaces of exploration and adventure that have in the past been associated with colonial-era storytelling, as well as the expansive evolutionism encountered in the works of H G Wells and others that considered the biological and ethical consequences of human presence in outer space.

Critics have suggested that Clarke's fiction can be divided into two stages: his earlier, 'genre sf' which suffered from 'wooden prose' but found expression through the depiction of science; and later works which introduce more speculative themes such as encounters with alien intelligence.[83] As outlined in Chapter 2, the appeal of science-fiction has been understood to be based on a sense of estrangement which is 'bound within spatial metaphors'.[84] Clarke's early works introduce spaceflight as such a spatial metaphor, particularly in three novels which were later incorporated into a 'space trilogy'; *Islands in the Sky* (1954), *Earthlight* (1955) and *The Sands of Mars* (1951). These novels, and Clarke's reputation in general, remain attached to 'a vision of the future which *assumed*, years before it happened, that space travel was both possible and desirable', while exemplifying Clarke's identity as 'a technological determinist, as well as an incurable optimist'.[85] Pertinent to this vision is the interplay between two different cultural *milieux*, represented by Clarke the science-fiction writer and Clarke the active BIS member, through which languages and narratives of interplanetarism were processed and disseminated.

Drawing together some of these themes, *Islands in the Sky* recounts the adventures of a sixteen-year-old 'space mad' boy who wins a competition to be sent into outer space.[86] The novel's setting depicts what was perceived at the time as an achievable goal of space exploration, the orbital space station. Indeed, technical specifications for 'orbital bases' had been published by the BIS in 1949, which envisioned a manned space station consisting of a large solar power dish along with living quarters, laboratories and workshops for a crew of twenty-four.[87] This vision is extrapolated by Clarke in *Islands in the Sky,* which is set among a series of space stations, including the Inner Station (500 miles from Earth), Meteorological Stations (6,000 miles), Biology Labs and Space Hospital (15,000 miles), and finally the Relay Stations (in a 'fixed' orbit of 22,000 miles). Clarke's 'relay stations' also drew from his own technological vision, published in 1945 in the amateur radio magazine *Wireless World,* often cited as the very first published conception of geostationary satellites, and this interface between theoretically-plausible space science and imaginative discourse was to form a cornerstone of Clarke's writing throughout his career.[88]

The literary connections between Clarke and the BIS are extended further into interplanetary space in *Earthlight* and *The Sands of Mars.* The latter sees a science-fiction writer transported to Mars on the space ship 'Ares' in what has been described as a 'then-revolutionary anti-Romantic' narrative which imagines life on Mars as 'more tedious than exotic'.[89] The protagonist's adventures on Mars relate to the on-going colonisation of the planet, including the establishment of a frontier settlement, and the discovery of basic life forms. Clarke's representation of the Martian landscape, like his description of space stations, appears to be influenced by discourses within the BIS. As early as January 1948, Mars was the subject of speculation as a credible destination for human spaceflight:

> When the ice of interplanetary travel has been broken and the majority of people have recovered from their surprise at the successful return of a Lunar Expedition, then will man's questing spirit turn to the first of the planetary targets. Of the two nearest – Venus and Mars – it is probable that Mars [...] will be chosen.[90]

There is little doubt that Clarke would have been aware of this article, as he was a member of the BIS Council at this time, and it is reasonable to suggest that aspects of *The Sands of Mars* were extrapolated from descriptions in this *JBIS* article. Based on observations conducted by the British Astronomical Association, Max W Wholey goes on to speculate about the surface conditions of Mars, including the conviction that a 'green belt' existed, which would likely be inhabited by 'some types of vegetation, most probably of a degenerate character'.[91] Sure enough, Clarke's fictitious exploration of the Martian surface incorporates a description of such plant life:

> They were not really very exciting [...] those around him now seemed to be made of sheets of brilliant green parchment, very thin but very tough [...] Quite a triumph of evolution.[92]

In this way, Clarke described a 'thoroughly realistic colonised Mars', based on the optimal scientific information available to him at the time, and mediated through the BIS.[93] He further extrapolates this 'knowledge' in the narrative by suggesting that cultivation of the plants in greater quantities would enable the generation of a breathable atmosphere for Mars, thinking through some of the problems associated with establishing human life on other planets.

Earthlight is an espionage thriller set two hundred years in the future, in which people have started to colonise the planets of the Solar System. The narrative takes place in various colonies on the Moon, and centres on a civil servant who is sent on a mission from Earth to uncover a spy who is working for a breakaway Federation of planetary colonies. The novel's narrative climax witnesses a battle between a domed fortress on the Moon and a Federation space cruiser. This encounter is said to have 'ended the domination of Earth and marked the coming of age of the planets', which Clarke compares to the way in which 'the American colonies turned against their motherland'.[94] With this broader scope, *Earthlight* is perhaps the most ambitious of Clarke's early works, and alongside the overriding theme of interplanetary colonies developing independently from Earth, a notable aspect of this novel is the representation of astronomy as an integral component of interplanetary science. This is highlighted in the placement of a Lunar observatory at the centre of the narrative in *Earthlight*, where 'the largest telescope ever built by man' is operated by a team of professional astronomers.[95] The significance of this telescope is not fully revealed until the epilogue, where it transpires that the spy had used it to send messages across space in Morse code. Moreover, Clarke assigns great importance to the development of astronomy in interplanetary space:

> Freed at last from the imprisoning atmosphere of Earth, astronomy had made giant strides; and indeed there was barely a branch of science that had not benefited from the lunar observatories.[96]

This projection into outer space brings astronomy's search for optimal viewing conditions to its zenith through the complete removal of atmospheric interference. In this passage, Clarke alludes to 'predictions' about astronomy in space, that he himself announced in a paper delivered to the BIS in December 1946:

> The most obvious and direct result of the crossing of space will be a revolution in almost all branches of science [...] An observatory on the Moon [...] would be many times as effective as one on Earth.[97]

This idea of astronomy in outer space was expanded upon three years later in the *BIS Journal* by the Cambridge astronomer Michael Ovenden, who discusses the possibility of setting up a lunar observatory, before describing in detail its anticipated benefits, which would include advancing the science of astronomy 'from observation to experiment'.[98] In these examples, Clarke's science-fiction drew on ideas that were engendered in the forums of the BIS, demonstrating how Clarke bridged the divide between his work as 'a writer and as a spaceflight advocate'.[99] The ways in which three key ideas – the Earth satellite, the Martian colony and the Lunar observatory – are inserted into Clarke's fiction further illustrate the close relationship between imaginative and technological conceptions of outer space in this context. Far from being incompatible, these discourses were complimentary, and created new intellectual spaces in which BIS members, science-fiction writers and readers explored the interplanetary idea in post-war Britain.

The language of Clarke's science-fiction is an important aspect to consider in terms of how he translated into fiction the technological progress in outer space that he envisioned, in a manner that resonated with Patrick Moore's goals in the popularisation of astronomy. One of the ways in which he sought to achieve this was through the adoption of maritime terminology. By the time Clarke was writing, this was an established technique in science-fiction, with the term 'space ship' first attributed in a reference to Jules Verne as early as 1880.[100] Clarke expanded on this framework to employ all manner of nautical terminology, based around the metaphor of space stations as inhabitable 'islands' in a sea of space. In one extract portraying an old-fashioned rescue mission, a space ship's 'crew' are even expected to conform to maritime ways:

> The traditions of space were as strict as those of the sea. Five men could leave the *Acheron* alive – but her captain would not be among them.[101]

Although this was an established trope in international science-fiction, there is evidence in some of Clarke's other writings of a particularly British type of naval heritage that was expected to inspire the exploration of space. In one extract from his early novel *Prelude to Space*, Clarke invokes an Elizabethan age of adventure:

> The line that stretched back to Drake and Raleigh and yet earlier voyages was still unbroken: only the scale of things had changed.[102]

Whereas nautical language is used by Clarke to evoke a sense of adventure at sea that would be familiar to readers, it also effectively endorses the British imperial histories associated with such language, and a sense in which human history is driven by the challenge of empire, a tendency that has been recognised in Clarke's work by various scholars.[103] While the connection between adventure and imperialism has been well-established, the translation of maritime adventure to outer space is a unique trope of science fiction, and brings with it its own set of specific geographical implications.

Over and above these historical maritime associations, a more personal meaning can be read into Clarke's use of the oceanic metaphor, which relates to the condition of weightlessness. In a chapter of *Islands in the Sky* triumphantly entitled 'Goodbye to Gravity', Clarke describes moving in zero gravity as 'rather like learning to swim underwater'.[104] Accordingly, being underwater represented for Clarke;

> a cheap and simple way of imitating one of the most magical aspects of spaceflight – weightlessness[105]

and one of the main reasons why Clarke started living in Ceylon (now Sri Lanka) in 1956 was so that he could spend more time practising his new hobby, scuba-diving.[106] A tall and ungainly figure, Clarke felt encumbered by what he thought of as the scourge of gravity. This feeling took on a heightened significance after 1962, when he suffered a severe attack of polio-induced paralysis, which continued to afflict him later in life through post-polio syndrome. Writing after this episode, his descriptions of weightlessness – or, lack of it – took on a more despondent tone:

> All our lives, we creatures of the land must drag the weight of our bodies around with us, envying the freedom of the birds and clouds.[107]

Even before his illness, Clarke considered being underwater and, ultimately, being in outer space, as a kind of salvation from the constant oppression by gravity on Earth.[108] Furthermore, by introducing characters with physical ailments who were able to thrive in lower-gravity environments, Clarke promoted a type of heroic identity that excelled through mental achievement rather than through physical prowess. This reflected his own identity as a scientific thinker, and as someone whose own body seemed at times to be a burden.

Clarke framed the relationship between people, science and gravity in an evolutionary context, noting how the human race had developed from creatures that once crawled out of the sea, and would eventually migrate to outer space in a return to a weightless environment, ultimately evolving into a different kind of creature, one that would be destined to exist in outer space. As Kilgore has explained, Clarke believed that 'the continued evolution of civilization depends on the conquest of space', and that 'human history [is] the expression of a biological imperative towards perfection'.[109] This perspective on evolutionism has implications for the geographies of outer space, with the lifting of what Clarke saw as Earthly environmental restrictions enabling humanity to actively take new directions in its own evolution. This continues some of the themes initially established in the works of H G Wells, who was among the first to imagine a diversity of intelligent life in space, and Olaf Stapledon, who applied this evolutionary framework to humankind in his imaginative visions of outer-space futures. As the examples in this section have demonstrated, Clarke wanted to bring these perspectives within a foreseeable future for humanity in space, by understanding space exploration as the next stage in the development of human history.

Conclusion: diverse cultures of outer space in post-war Britain

This chapter's selected examples have served to demonstrate the ways in which contrasting discourses of local, regional and national identity have characterised popular cultures of outer space in post-war Britain. While some figures in outer-space culture attempted to legitimise the science of outer space and translate it for broad public audiences, alternative cultures of outer space also thrived, no less popular, but more attuned to concepts of the unknown and the mythical in their understandings of outer space. So, whereas Arthur C Clarke used up-to-date scientific knowledge gleaned from the BIS in his future-realist space fiction, deploying a language of maritime metaphors and evolutionary development, Nigel Kneale, drawing from folk cultures of storytelling and a strong social grounding, brought to bear darker and more ambiguous understandings of outer space to his genre-defining teleplays. Similarly, Patrick Moore's rejection (at least in his professional persona) of 'space-borne crockery' excluded the possibility of extra-terrestrial visitations to Earth in his domestication of post-war space science, a prospect that was embraced by the emergent popular culture of the UFO, whose observational practices firmly embedded it within broader cultures of outer space. This diversity of outer-space popular culture is evident along a spectrum from active 'interplanetarism' on one end, to the mythical sublime on the other, incorporating cultures that were representational and performative, practical and observational, fictional and instructive.

Together, the case studies in this chapter have sought to envision humankind's active role in the understanding and possible colonisation of outer space, while also appreciating the unknowable, awe-inspiring and terrifying aspects of outer space. In comparison to the Astrofuturist mode of engagement with outer space in the post-war period, which came to prominence in America, these British engagements with outer space, while also encompassing a marriage of scientific and imaginative approaches across a range of media, have tended to demonstrate a more diverse set of cultural understandings. These differences can partly be explained by a contrasting national experience of the Second World War, the social and cultural imprints of empire and postcolonialism, and an appreciation of alternate traditions to modernism in the twentieth century. A geographical approach has provided a window on the characteristics of these outer-space imaginaries, from the spaces of aerial observation to the expansive imaginative spaces of evolutionary futures. The next chapter focusses on the scale of the international and the global in outer-space culture, as the practical and political aspects to space science in Britain caught up with the imaginative endeavours of the early-post-war period.

Notes

1 Fraser MacDonald, 'Instruments of Science and War: Frank Malina and the Object of Rocketry', in *Geography, Technology and Instruments of Exploration*, ed. by Fraser MacDonald and Charles W J Withers (Farnham: Ashgate, 2015), 219–240 (p.232).

2 De Witt Douglas Kilgore, *Astrofuturism: Science, Race and Visions of Utopia in Space* (Philadelphia: University of Pennsylvania Press, 2003).

3 John R Cook, 'Adapting Telefantasy – the "Doctor Who and the Daleks" Films', in *British Science Fiction Cinema*, ed. by Ian Q Hunter (London: Routledge, 1999), 113–127; John Tulloch, 'Producing the National Imaginary –

Doctor Who, Text and Genre', in *A Necessary Fantasy? The Heroic Figure in Children's Popular Culture*, ed. by Dudley Jones and Tony Watkins (London: Garland, 2000), 363–394.

4 Dave Rolinson, and Nick Cooper, '"Bring Something Back" – the Strange Career of Professor Bernard Quatermass', *Journal of Popular Film and Television*, 30 (2002), 158–165 (p.159); James Chapman, '*Quatermass* and the Origins of British Television Sf', in *British Science Fiction Television – a Hitchhiker's Guide*, ed. by John R Cook and Peter Wright (London: I B Tauris, 2006), 21–51.

5 Fraser MacDonald, 'High Empire: Rocketry and the Popular Geopolitics of Space Exploration, 1944–62', in *New Spaces of Exploration – Geographies of Discovery in the Twentieth Century*, ed. by Simon Naylor and James Ryan (London: I B Tauris, 2010), 196–221.

6 Andy Murray, *Into the Unknown – the Fantastic Life of Nigel Kneale* (London: Headpress, 2017), p.11.

7 Fiona Angwin, *Manx Folk Tales* (Stroud: The History Press, 2015).

8 Murray, *Into the Unknown* (p.10).

9 Paul Wells, 'Apocalypse Then! The Ultimate Monstrosity and Strange Things on the Coast ... An Interview with Nigel Kneale', in *British Science Fiction Cinema*, ed. by Ian Q Hunter (London: Routledge, 2002), 48–56 (p.52).

10 Murray, *Into the Unknown* (p.45).

11 Murray, *Into the Unknown* (p.42).

12 Rolinson and Cooper, 'Bring Something Back'.

13 Wells, 'Apocalypse then!' (p.56).

14 Stephen Bissette, 'The Quatermass Conception', in *We Are the Martians – the Legacy of Nigel Kneale*, ed. by Neil Snowdon (Hornsea: Electric Dreamhouse, 2017), 95–152.

15 Murray, *Into the Unknown*.

16 Mark Chadbourn, 'The King of Hauntology', in *We Are the Martians – the Legacy of Nigel Kneale*, ed. by Neil Snowdon (Hornsea: Electric Dreamhouse, 2017), 11–24 (p.15).

17 Rolinson and Cooper, 'Bring Something Back' (p.159); Angus Calder, *The Myth of the Blitz* (London: Pimlico, 1991).

18 Wells, 'Apocalypse then!' (p.53).

19 'Quatermass and the Pit', *The Times*, 23rd December 1958, 54341 (1958), 3.

20 Jason Jacobs, *The Intimate Screen: Early British Television Drama* (Oxford: Oxford University Press, 2000), p.137.

21 Kilgore, *Astrofuturism*.

22 'Coloured Leaders Criticize B.B.C.', *The Times*, 24th December 1958, 54342 (1958), 4.

23 Chadbourn, 'The King of Hauntology' (p.23).

24 *We Are the Martians – the Legacy of Nigel Kneale*, ed. by Neil Snowdon (Hornsea: Electric Dreamhouse, 2017).

25 Wells, 'Apocalypse then!' (p.55).

26 John Montgomery, *The Fifties* (Leicester: Blackfriars Press, 1965), p.78.

27 William J Dewan, '"A Saucerful of Secrets": An Interdisciplinary Analysis of UFO Experiences', *The Journal of American Folklore*, 119 (2006), 184–202.

28 Alexander Geppert, 'Extraterrestrial Encounters: UFOs, Science and the Quest for Transcendence, 1947–1972', *History and Technology*, 28 (2012), 335–362 (p.336).

29 James Miller, 'Seeing the Future of Civilization in the Skies of Quarouble: UFO Encounters and the Problem of Empire in Postwar France', in *Imagining Outer Space*, ed. by Alexander Geppert (Basingstoke: Palgrave Macmillan, 2012), 245–264.

30 Greg Eghigian, 'Making UFOs Make Sense: Ufology, Science, and the History of Their Mutual Mistrust', *Public Understanding of Science,* 26 (2017), 612–626.; Pierre Lagrange, 'A Ghost in the Machine: How Sociology Tried to Explain (Away) American Flying Saucers and European Ghost Rockets, 1946–47', in *Imagining Outer Space*, ed. by Alexander Geppert (Basingstoke: Palgrave Macmillan, 2012), 224–244.

31 Carl Jung, *Flying Saucers – A Modern Myth of Things Seen in the Skies* (London: Routledge & Kegan Paul, 1959), p.xiv.

32 Christopher Partridge, 'Haunted Culture – the Persistence of Belief in the Paranormal', in *The Ashgate Research Companion to Paranormal Cultures*, ed. by Olu Jenzen and Sally R Munt (Farnham: Ashgate, 2013), 39–50 (p.39).

33 David Clarke, 'Extraordinary Experiences with UFOs', in *The Ashgate Research Companion to Paranormal Cultures*, ed. by Olu Jenzen and Sally R Munt (Farnham: Ashgate, 2013), 79–94.

34 David Clarke, *The UFO Files* (London: Bloomsbury, 2012).

35 Lagrange, 'A Ghost in the Machine' (p.230, p.232).

36 Geppert, 'Extraterrestrial encounters' (p.335).

37 Clarke, 'Extraordinary Experiences' (p.80).

38 Clarke, *The UFO Files.*

39 Kew, London, National Archives, DEFE – Records of the Ministry of Defence, DEFE 44/119, DSI-JTIC Report No. 7.

40 DEFE – Records of the Ministry of Defence, DEFE 44/119, DSI-JTIC Report No. 7.

41 DEFE – Records of the Ministry of Defence, DEFE 44/119, DSI-JTIC Report No. 7.

42 Peter Adey, '"Ten Thousand Lads with Shining Eyes Are Dreaming and Their Dreams Are Wings": Affect, Airmindedness and the Birth of the Aerial Subject', *cultural geographies* 18 (2010), 63–89 (p.65).

43 James Chapman, 'Onward Christian Spacemen: Dan Dare – Pilot of the Future as British Cultural History', *Visual Culture in Britain,* 9 (2008), 55–79; Oliver Dunnett, 'Framing Landscape: Dan Dare, the Eagle and Post-War Culture in Britain', in *Comic Book Geographies*, ed. by Jason Dittmer (Stuttgart: Franz Steiner, 2014), 27–40.

44 David Edgerton, *England and the Aeroplane – an Essay on a Militant and Technological Nation* (London: Macmillan, 1991).

45 Lionel Beer, *An Anecdotal History of BUFORA* (1983 [Reprinted October 2003]) https://bufora.org.uk/documents/19752003HistoryofBUFORALionel-Beer.pdf [5th August 2019].

46 Charles A Stickland, 'Aims', *BUFORA Journal and Bulletin*, 1, no.1 (Summer 1964), 1.

47 G G Doel, 'Home Reports', *BUFORA Journal and Bulletin*, 1, no.1 (Summer 1964), 5–6 (p.5).

48 J Cleary-Baker, 'Home Reports', *BUFORA Journal and Bulletin*, 1, no.2 (Autumn 1964), 10–12 (p.12).

49 E Conrad Miller, and J L Smith, 'Some Considerations Regarding the Possibility of Contact with Intelligent Extra-Terrestrial Beings', *BUFORA Journal and Bulletin*, 1, no.3 (Winter 1964), 4–7.

50 Norman Oliver, 'The Winter Skies', *BUFORA Journal*, 2, no.3 (Winter 1967/8), 3–4 (p.3).

51 J Cleary-Baker, 'Editorial', *BUFORA Journal*, 2, No.11 (1970), 2–3 (p.3).

52 Partridge, 'Haunted Culture'.

53 Geppert, 'Extraterrestrial encounters' (p.335, p.341).

54 Peter Bowler, 'Parallel Prophecies: Science Fiction and Futurology in the Twentieth Century', *Osiris,* 34 (2019), 121–138 (p.132).

55 Chris Dunkley and Clive Cookson, 'Obituary: Sir Patrick Moore, Astronomer', *Financial Times*, 9th December 2012.
56 Patrick Moore, 'Editorial', *Spaceflight,* 1 (1956), 1 (p.1).
57 Reading, Caversham Park, BBC Written Archives Centre, Correspondence, Paul Johnstone to Patrick Moore, 22nd June 1956 ('Patrick Moore correspondence / Talks 1: 1954-1957').
58 Frederick C Durant III, 'A Message from F C Durant III, President of the IAF', *Spaceflight* 1 (1956), 2 (p.2).
59 Patrick Moore, 'Editorial', *Spaceflight*, 1 (1956), 1 (p.1).
60 Author's interview with Patrick Moore, 2nd November 2009, Selsey, West Sussex.
61 Patrick Moore, 'Editorial', *Spaceflight*, 1 (1957), 45–46 (p.45).
62 Gordon V E Thompson, 'The British Interplanetary Society, 1933–1979', *Spaceflight*, 21 (1979), 634–658.
63 Vincent C Reddish, 'The First Days of Sputnik 1', *Spaceflight*, 1 (1958), 198.
64 Rip Bulkley, 'Harbingers of Sputnik: The Amateur Radio Preparations in the Soviet Union', *History and Technology*, 16 (1999), 67–102.
65 W Patrick McCray, 'Amateur Scientists, the International Geophysical Year, and the Ambitions of Fred Whipple', *Isis,* 97 (2006), 634–658.
66 Denis Cosgrove, *Apollo's Eye: A Cartographic Genealogy of the Earth in the Western Imagination* (Baltimore: John Hopkins University Press, 2001).
67 Reading, Caversham Park, BBC Written Archives Centre, Patrick Moore correspondence, Talks 1: 1954–1957, Patrick Moore to Paul Johnstone, 17th October 1956.
68 Timothy Boon, *Films of Fact: A History of Science in Documentary Films and Television* (London: Wallflower, 2008), p.211.
69 Reading, Caversham Park, BBC Written Archives Centre, VR/57/229, 'An Audience Research Report, 24th April 1957'.
70 Reading, Caversham Park, BBC Written Archives Centre, Production notes, *The Sky at Night*, 24th April 1957 [on microfilm].
71 'An Audience Research Report, 24th April 1957'.
72 London, Fitzrovia, British Film Institute National Archive, 'Ten Years of Astronomy', 28th April 1967 [on VHS tape].
73 Reading, Caversham Park, BBC Written Archives Centre, Production notes, *The Sky at Night*, 20th June 1957 [on microfilm].
74 Reading, Caversham Park, BBC Written Archives Centre, Production notes, *The Sky at Night*, 14th December 1957 [on microfilm].
75 Reading, Caversham Park, BBC Written Archives Centre, Patrick Moore correspondence, Talks 1: 1954–1957, Patrick Moore to Paul Johnstone, 12th November 1956.
76 Fitzovia, London, British Film Institute National Archive, 'Sky at Night Excerpts', 17th September 1958 [on DVD].
77 Author's interview with Patrick Moore, 2nd November 2009, Selsey, West Sussex.
78 BBC, 'The Sky at Night – 1963 Bases on the Moon', *BBC The Sky at Night* https://www.bbc.co.uk/programmes/p00m72mf [20th August 2020].
79 'Ten Years of Astronomy', 28th April 1967.
80 Caversham Park, Reading, BBC Written Archives Centre, TV Travel and Features, Sky at Night 1957-1969, Julia Gaitskell to Jean Baxter, 30th April 1965, T50/88.
81 Christopher Allan, and Steuart Campbell, 'Flying Saucer from Moore's?', *Magnolia,* 23 (1986); Geppert, 'Extraterrestrial encounters'.
82 Caversham Park, Reading, BBC Written Archives Centre, Production notes, Film no. 13, 'First Hand 3: Flying Saucers', 4th December 1956, T32/620/2 [on microfilm].

83 Peter Nicholls, 'Clarke, Arthur C', in *The Encyclopedia of Science Fiction*, ed. by P Nicholls (London: Granada, 1981), 121–122.

84 Rob Kitchin and James Kneale, 'Science Fiction or Future Fact? Exploring Imaginative Geographies of the New Millennium', *Progress in Human Geography,* 25 (2001), 19–35 (p.21).

85 Andy Sawyer, 'Sir Arthur C[harles] Clarke (1917–2008)', in *Fifty Key Figures in Science Fiction*, ed. by Mark Bould, Andrew M Butler and Sherryl Vint (New York: Routledge, 2009), p.55 [emphasis in original]; Albert I Berger, 'Science-Fiction Critiques of the American Space Program', *Science Fiction Studies*, 15 (1978), n.p.

86 Arthur C Clarke, *Islands in the Sky* (Harmondsworth: Puffin, 1972 [1954]).

87 Harry E Ross, 'Orbital Bases', *Journal of the British Interplanetary Society,* 8 (1949), 1–19.

88 Arthur C Clarke, 'Extra-Terrestrial Relays – Can Rocket Stations Give World-Wide Radio Coverage?', *Wireless World* (1945), 305–308.

89 Robert Crossley, *Imagining Mars: A Literary History* (Middletown CT: Wesleyan University Press, 2011), p.208, p.263.

90 Max W Wholey, 'Conditions on the Surface of Mars', *Journal of the British Interplanetary Society,* 7 (1948), 2–20 (p.2).

91 Wholey, 'Conditions on the Surface of Mars' (p.8).

92 Arthur C Clarke, *The Sands of Mars* (London: Sidgwick and Jackson, 1976 [1951]), p.94–95.

93 Adam Roberts, *The History of Science Fiction* (New York: Macmillan, 2005), p.212.

94 Arthur C Clarke, *Earthlight* (London: Pan, 1963 [1955]), p.146.

95 Clarke, *Earthlight* (p.19).

96 Clarke, *Earthlight* (p.32).

97 Clarke, 'The Challenge of the Space Ship' (p.71).

98 Michael W Ovenden, 'Astronomy and Astronautics', *Journal of the British Interplanetary Society,* 8 (1949), 180–193.

99 Kilgore, *Astrofuturism* (p.111).

100 Oxford English Dictionary, 'astronautics, n.' *Oxford English Dictionary* (2020) https://www.oed.com/view/Entry/12275 [10th August 2020].

101 Clarke, *Earthlight* (p.140).

102 Arthur C Clarke, *Prelude to Space* (London: Sidgwick and Jackson Ltd., 1953), p.11.

103 Kilgore, *Astrofuturism*; Robert Poole, 'The Challenge of the Spaceship: Arthur C Clarke and the History of the Future, 1930–1970', *History and Technology,* 28 (2012), 255–280.

104 Clarke, *Earthlight* (p.105).

105 Neil McAleer, *Odyssey: The Authorised Biography of Arthur C Clarke* (London: Victor Gollancz 1992), p.105.

106 Oliver Dunnett, 'Imperialism, Technology and Tropicality in Arthur C Clarke's Geopolitics of Outer Space', *Geopolitics* (2019) [online].

107 Arthur C Clarke, *Report on Planet Three and Other Speculations* (London: Pan Books, 1984), p.226.

108 Thore Bjørnvig, 'Transcendence of Gravity – Arthur C Clarke and the Apocalypse of Weightlessness', in *Imagining Outer Space – European Astroculture in the Twentieth Century*, ed. by Alexander Geppert (New York: Palgrave Macmillan, 2012), 127–146.

109 Kilgore, *Astrofuturism* (p.115).

5   The British space programme

Geopolitics and empire

This chapter broadens perspectives in the post-war period by consider-ing the ways in which geopolitical discourse became integral to the for-mulation of a complex and multi-faceted British culture of outer space. The geopolitics of outer space in the post-war period is usually associ-ated with narratives involving the legacies of German wartime rocketry, the capture of iconic military hardware and talismanic scientific experts by American and Soviet forces at the end of the Second World War, and the subsequent 'space race' between the two world superpowers, as they vied for supremacy in the heavens.[1] Such accounts act as a tem-plate for dominant narratives of space exploration, often in terms asso-ciated with a bi-polar understanding of the Cold War and written from the perspectives of hegemonic cultural and political authority. Looking at the British context, however, allows an alternative story to emerge, involving an equally significant set of changes in world power dynam-ics. These include the complex and contingent shifts from colonialism to postcolonialism, implicating a changing set of dynamics between cultures of nationalism and internationalism. That these narratives and cultural histories have remained, for the most part, untold, can perhaps be traced to the perceived failure of a British national space programme. That is not to say, however, that geopolitical cultures of outer space in Britain did not exist or had no broader impact. Indeed, the British Interplanetary Society remained central to geopolitical cultures of outer space in Britain, fostering utopian visions of spaceflight, promoting international collaborations in outer space technology, and speculating about imagined future launch sites of a British space programme. This provides a compelling insight into aspects of British cultural and polit-ical identity on the world stage in the post-war period, while also help-ing to understand the cultural underpinning of the limited British space programme of the 1960s. Following a discussion of the ways in which researchers have theorised the geopolitics of outer space, this chapter turns to the geopolitical narratives associated with British outer-space activities in the post-war period.

Outer space and critical astropolitics

Scholars of geopolitics both within and outside the discipline of geography have become increasingly aware of the need to consider outer space as a key analytical realm. Indeed, despite the widespread adoption of the 1967 UN Outer Space Treaty, that decreed outer space as a non-sovereign, shared international space, there has been a general acceptance that Earth-orbital spaces have become militarised, alongside an acknowledgement that spaceflight is inherently linked to the inter-continental ballistic missile technologies of the Cold War.[2] Such understandings have led some geopolitical thinkers to adopt neo-classical models in explaining and promoting state involvement in spaceflight. As such, 'spacepower theorist' Everett C Dolman has advocated an aggressive US policy in outer space, drawing heavily on classical theorists Halford Mackinder (1861–1947) and Alfred T Mahan (1840–1914), going so far as to urge the United States to 'seize military control of the Low Earth Orbit'.[3] Furthermore, a wide range of studies advocating a neo-classical geopolitics of outer space has recently emerged in specialist space policy journals, with a particular tendency to promote the application of Mahanian sea-power theories from the nineteenth century to the militarised realm of outer space, in interventions typically saturated with nostalgia for a 'lost age' of US space dominance.[4] Whereas Dolman and others clearly are in favour of taking action to support a neo-liberal agenda of space domination, additional studies have favoured 'neo-classical astropolitics' as an explainer of activities in outer space since the mid-twentieth century. One such study, although providing a welcome focus on European space policy, defines geopolitics as 'a dynamic struggle among strong states who seek to seize new "space" and organise it to fit their own interests', contending that such state and supra-state organisations frame our understandings of spaceflight in the modern era.[5] Interventions along these lines typically examine state programmes, such as US space policy under President George W Bush, or more widespread patterns in the securitisation of outer space, and although occasionally critical of such programmes, tend to be theoretically grounded in mainstream international relations literature or neo-classical approaches to geopolitical thought.[6]

Running counter to this school of thought in 'astropolitics' has emerged a critical geopolitics of outer space, led by authors including historical geographer Fraser MacDonald, who has argued against the proliferation of 'undead' neo-classical models in thinking about and promoting the neo-liberal domination of outer space.[7] While acknowledging that outer space plays an integral role in the modern lives of citizens in most developed countries, MacDonald argues that geopolitics and spaceflight are intrinsically linked and that 'the colonisation of space, rather than being a decisive and transcendent break from the past, is merely an extension of longstanding regimes of power'.[8] However, rather than adopting neo-classical theories in explaining the relevance of these 'extensions of power', this work looks to the emergence of critical geopolitics. This contends that 'geopolitics should be re-conceptualised as a discursive practise', as opposed to a normative set of theories applicable to 'real-world' situations through state strategies, as one might characterise the above iterations of 'neo-classical astropolitics'.[9] In other words, critical astropolitics, in the same ways as critical geopolitics has successfully argued, can be understood as part of 'a refusal to accept the abstract logic' of established classical theory, 'but instead embody it in historically and culturally specific interests'.[10] Fully understanding the geopolitics of outer space therefore involves engaging critically with, firstly, the granular details of representation and practise that form part of such discourses and second, the broader narratives of national, (post-)colonial and international identity formation, as co-constitutive elements of astropolitics. In this way, an appreciation of the broader geopolitical *cultures* of outer space, as opposed to a narrow reading of space policy in relation to particular state interests, offers a more representative and critically rigorous basis for understanding the geopolitics of outer space.

Certain works in this form of critical astropolitics have examined the construction of nationalism and colonialism in outer space discourses, typically focussing on American and European case studies. Here, research into the US space programme has identified how nationalistic cultures of exploration, technological determinism and religiosity were fundamental to American astropolitics in the twentieth century.[11] Cultural works such as those of space artist Chesley Bonestell (1888–1986) have been foregrounded, his speculative paintings of colonised Lunar landscapes having drawn influence from the nineteenth-century frontier tradition of sublime landscape painting.[12] This cultural-political thread of American Manifest Destiny in outer space has been picked up in the late twentieth century, with media reports of NASA's *Mars Pathfinder* missions of the late 1990s exhibiting distinctively colonial subtexts. The discourse surrounding these missions; including the progress of science in outer space, the Westernised nomenclature of topographic features on Mars and the use of Earth analogues to understand and know the Martian landscape, has been shown to demonstrate how Mars was 'constructed [...] as a place to be colonized' by humans in prospective future missions to the Red Planet.[13] Similar examples of colonialism in the language of understanding outer space have been explored in studies on nineteenth-century astronomers of Mars and on the production of nature in Kim Stanley Robinson's science-fictional 'Mars trilogy'.[14] These examples all demonstrate how colonialism has been used as a metaphor for understanding spaceflight in the twentieth century, particularly in the American context, and provide insights into how such an approach might help to understand geopolitical cultures of outer space in broader contexts.

Further productive lines of enquiry in critical analysis of the geopolitical and cultural aspects of space programmes have taken a postcolonial approach. As such, the Kourou rocket range in French Guiana has been interpreted as a 'fertile field of representation', through which layers of meaning around the French and European space programmes can be excavated.[15] This site, originally established as the home of a national French space programme in 1968, was later co-opted by the European Space Agency for the successful *Ariane* space rocket. Here, the selection of the rocket site in French Guiana has been connected not only to the 'latitudinal boost' provided to space rockets by equatorial launching sites and the associated attainment of geosynchronous orbital positions in space, but also to deeply-embedded historical networks of colonialism, as the shortlist of French rocket sites reportedly 'parallel[ed] the list of potential penal colonies a century earlier'.[16] Furthermore, the connected settlement of Kourou has been characterised as 'a new colony [...] that reproduces an old colonialism in its racial divides reinforced by social class'.[17] This tension was of course replicated on a wider symbolic scale in the somewhat incongruous launching of European rockets and satellites from this location in South America. This postcolonial critique has also been applied in broader theoretical terms, with one critic explaining how 'the notion of a biologically-engrained need of humans to conquer new horizons is appealed to by the European Space Agency' in its colonialist-inflected public-facing discourse.[18] This point of view is supported by further analysis of ESA promotional material for the *Centre Spatial Guyanais*, which draws on the achievements of classical European civilisation, universal human development and genealogies of scientific progress.[19] In offering detailed accounts of geopolitical cultures of outer space, what these more critical studies by MacDonald and others have in common is the understanding of a tension between the implied utopianism of modernity and the messier post-colonial or Cold War contexts in which such ideas evolved.

In the UK, the common themes of nationalism and colonialism in the post-war period became imbricated in the transition from Empire to Commonwealth and a changing relationship between the UK and its wartime allies the United States, the USSR and the nations of Western Europe. The importance of British cultural discourses involving the monarchy of Elizabeth II, post-war technological developments and narratives of exploration, that were embedded in the national psyche, are all relevant to the development of British geopolitical cultures of outer space. The remainder of this chapter considers the international connections that were forged by spaceflight promoters in the post-war period, shows how optimistic visions of internationalism in outer space became compromised by narrations of empire and nationalism, and explores the geopolitical 'pivots' associated with the nascent British space programme.

British internationalism in outer-space research

In 1946, members of the British Interplanetary Society were able to re-group and take stock of wartime developments in rocketry and their likely effect on the advancement of spaceflight. With the help of this post-war rocketry stimulus, the Society started to broaden its horizons once more, with the internationalist outlook that was promoted in the 1930s retaining a significant role. However, the changing international context of this period tended to temper the outright optimism of the Society's pre-war outlook, towards a version of internationalism that would be contingent on British leadership and shaped by new geopolitical concerns.

At a meeting in October 1947, Arthur 'Val' Cleaver, a key player in the post-war BIS who later went on to become Chief Rocket Propulsion Engineer at Rolls-Royce, presented a paper to the BIS entitled 'The Interplanetary Project', in which he set out two possibilities for the future of spaceflight.[20] The first he called 'the utopian view', whereby a post-warfare global society would come together with the advent of new technologies to explore interplanetary space for the good of all humankind, a vision that drew heavily from the Society's pre-war idealist internationalism (see Chapter 3). Cleaver's second possibility, however, was that militaristic and nationalistic motives would spur the development of spaceflight, a vision borne out of the calamities of the Second World War and the rising tensions of the Cold War, and a warning to BIS members on the likely cost of failure of internationalism in spaceflight research. BIS members Gordon Thompson and Les Shepherd later suggested that this anxiety was the principal reason for the setting up of further links between spaceflight societies in this period, which culminated with the establishment of the International Astronautical Federation (IAF) in 1951.[21]

Now labelled 'the world's leading space advocacy body', the IAF was first conceived in correspondence between the BIS, the *Groupement Astronautique Français* and the *Gesselschaft für Weltraumforschung* (*GfW*).[22] BIS members were said to have been 'strongly in favour of the new international federation of astronautical societies, providing that the autonomy of the existing national groups was preserved'.[23] Immediately apparent is a sense of the tensions between the promise of internationalism and the paradoxical need for autonomy from any overriding authority that such an organisation might threaten. Nonetheless, a preliminary meeting was organised in Paris on the 30th September 1950, at the Grand Amphitheatre of the Sorbonne, with a reported turnout of over one thousand delegates, while a smaller business meeting between the representatives of eight societies took place the next day at the French Aero Club, chaired by Val Cleaver.[24] At the latter it was decided that the IAF should be formally inaugurated at a 1951 Congress in London, which would be organised by the BIS. It was also reported that the American Rocket Society was not able to attend the Paris Congress and that;

> the non-participation of the Americans at Paris had been due to a feeling on their part that they were so far ahead of the rest of the world in rocket development that they had little to receive, only to give, from any project for international collaboration.[25]

It appears that the delegates at the Paris Congress were only vaguely aware of developments in American spaceflight and this commentary perhaps also hints at the influence of US authorities in protecting state secrets in rocketry, which at this time included the results of testing of captured V–2 rockets.[26] In any case, the American non-participation in this initial meeting represents the limits of internationalism in discourses of spaceflight research at this time and foreshadows the rivalries and consternation that became more typical in later Cold War discourse.

In a retrospective article, Paris BIS delegate Les Shepherd recounted the process of establishing the IAF. In his view,

> the proposed international body was envisaged as being a much more conservative federation of the various national societies, [although] many of the representatives hoped that it might eventually become more than this.[27]

Shepherd's is a realistic account of the role that the IAF would adopt, however, he also hints at a more ambitious view of the future role of the Federation, which for many BIS members involved its incorporation into the United Nations. Founded in 1945, the UN sought to promote international co-operation in a way that would have seemed promising to the BIS and it was deemed the natural home for global collaborative projects such as the International Map of the World.[28] Writing in the *BIS Journal*, Guenter Loeser addresses this prospect directly:

> As evolution can neither be stopped nor kept back, some day the United Nations, a world parliament or some other international institution will take up the problem, working for peaceful space travel as a common aim of civilisation.[29]

The contributions by Shepherd and Loeser hint at a sense of disappointment with what the IAF would be able to offer, as delegates looked forward to the possibility of 'loyal and unselfish international collaboration' under the auspices of the United Nations.[30] Nonetheless, the new Federation was largely seen as a success and in May 1951, 'the BIS was able to circulate a full draft Constitution for the IAF'.[31] This Constitution was approved at the London Congress, with the BIS presiding over four days of working sessions, alongside social occasions and a public meeting, with press attendance throughout.

The working sessions consisted of papers being delivered at Caxton Hall in Westminster, in English, French and German (with translated summaries), centred on the theme of 'The Earth-Satellite Vehicle', which was seen as 'the first and essential task in the conquest of space'.[32] Perhaps of equal significance as part of the 'sites [of] performance' at the Congress were the scheduled social events, which were said to have 'played a pleasant and important part in helping to forge international bonds of goodwill between the various delegates'.[33] As part of such activities, 'the BIS presented all delegates with tickets to the South Bank (Festival of Britain) exhibition and several BIS Council Members accompanied them on a visit there'.[34] This exhibition had opened in the summer of 1951 and the delegates would no doubt have gravitated towards the rocket-like Skylon tower and the Dome of Discovery with its Outer Space exhibit. In the Dome of Discovery, 'the body and outer space were constituted as the appropriate frontiers for discovery, rather than foreign lands' and such exhibits, alongside new architectural forms in the International Modernist style, may have encouraged the IAF delegates to see themselves as the new explorers of a modern age.[35]

Images from the German *GfW* (Society for Space Research) brochure of the London Congress demonstrate pictorially the key thematic elements of this event. The images from this brochure, whose authorship is somewhat unclear, can be seen alongside other sources such as the *BIS Journal* and the Congress programme of events, as emblematic of the broader 'mission statement' of the Congress, while its very existence echoes the earlier acts of translation that characterised the international dimensions of spaceflight research in this period. The first image from the brochure (Fig. 5.1) shows the whole Earth as seen from a point in space, an image whose transcendental qualities have endured for centuries in the Western imagination.[36] With Europe's hemisphere prominent, a UK flag is thrust into the blankness of outer space from a point in Northern Europe, from which, in turn, emanates a group of national flags. These are overlaid by a 'Dove of Peace' by Pablo Picasso, initially commissioned as an emblem of the First International Peace Conference in Paris in 1949. Another image from the last page of the brochure (Fig. 5.2) is similar, but more peculiar, incorporating a photomontage of the heads of the major characters in European spaceflight research, including Eugene Sanger, Hermann Oberth and Wernher Von Braun, being led towards the Moon under a giant hat belonging to 'London' by Val Cleaver, pictured sitting at a desk, ringing a bell. Two characters seem to be turned away by the anthropomorphic Moon, one of whom appears to be carrying an American flag and can be identified as the then US President Harry Truman. The other caricature, sitting astride a rocket, is likely to be Robert Goddard, the lone American rocketry pioneer. With their striking use of photomontage and collage techniques, these images bear the influence of the Berlin Dada art movement of the 1920s, whose avant-garde works were often intensely satirical and political, arising partly as a reaction to the horrors of the Frist World War and its associated national rivalries.[37] However, rather than acting as images of overt protest, the symbolic message implied by these two montages is that joint European co-operation, with British leadership, is the favoured means by which humankind should colonise outer space, as opposed to unilateral American or individualistic progress towards this goal. The prominence of Britain reflects London's status as the host city of the 1951 IAC, but this in turn is a result of the

Figure 5.1 Front cover of Society for Space Research (*GfW*) pamphlet for the 1951
International Astronautical Congress

Source Credit: British Interplanetary Society

instrumental role that the British organisers played in setting the agenda for
the first few years of the IAF.

Following subsequent Congresses at Stuttgart (1952), Zurich (1953) and
Innsbruck (1954), the IAF introduced its journal, *Astronautica Acta*, which
remains one of the leading journals in spaceflight research, while the IAC
still meets regularly to this day. As such, the establishment of the IAF in the
early 1950s, including its two opening Congresses, are important events to
consider, not only in the history of spaceflight research and the significant
role British representatives had in its development, but also in relation to the
ways in which the BIS formulated and articulated its geopolitical discourse
of spaceflight in the early-post-war period. This form of internationalism
was moulded by a sense of British leadership and the primacy of European
voices in the anticipated conquest of space, moving on from the more purely
internationalist agenda of the pre-war BIS. The significance of the *cultures*

Vergessen fallen . . ." So ge-
schrieben im Jahr 1951 und ge-
druckt in einer großen westdeut-
schen Wochenzeitschrift. Hoffent-
lich fällt der kuriose Beitrag
ebenfalls recht bald dem ewigen
Vergessen anheim.

**Zum
Londoner
Kongreß**

*„Ob sie alle unter einen Hut
kommen?"*

Seite 96   WELTRAUMFAHRT 1951, Heft 4

Figure 5.2 Last page of Society for Space Research (*GfW*) pamphlet for the 1951
         International Astronautical Congress. The captions translate as 'The
         London Congress'/'will they all come under one hat?'

Source Credit: British Interplanetary Society

of geopolitics has also been foregrounded in defining such notions of spaceflight, including social relations at international symposia, key characters
as icons in narratives of spaceflight research and the artistic representations
that helped shape such narratives.

New Elizabethanism and a Commonwealth space programme

From the mid-1950s onwards, the concept of a British space programme
slowly began to take shape in popular and institutional discourse, but was
always bound by geopolitical contingencies. While the IAF did not actually have the power to conduct serious collaborative research, the BIS was
well aware that the British government was not willing to act alone, having
'neither the practical will nor the resources to become involved in a space
race'.[38] So, maintaining the spirit of scientific internationalism, the BIS

in this period chose to promote collaboration between Commonwealth nations as the desired means of achieving spaceflight. This shift not only came about because of the need to share research and development costs, but also because of the perceived geographical advantages held by the Commonwealth as a whole. A 1961 article in *Spaceflight* by E D G Andrews highlighted these benefits:

> The British Commonwealth has one asset for a space programme possessed by no other single community in the world: its scatter. Moreover, member nations control territories at nearly every latitude between 50° south and the North Pole, with an area of Antarctica thrown in for good measure [...] The top of Mount Kenya [...] seems to be the ideal launching site for interplanetary probes.[39]

Here the Commonwealth is conceptually enrolled from pole to pole for the benefit of British spaceflight, connecting the past exploits of Empire to the future exploration of outer space. One of the most vocal proponents of this 'Commonwealth Space Project' was the aforementioned Gordon Thompson, one of the Society's most active post-war members. Thompson declared on behalf of the Society that 'the British Commonwealth should launch satellites and undertake space research', attempts which 'must not be mere imitations of American and Russian feats'.[40] With this in mind, the BIS organised a Commonwealth Spaceflight Symposium in London in August 1959. One presentation suggested 'the use of Antarctic territory as the Commonwealth satellite launching site for a pole-to-pole orbit', tacitly approving of the UK's claim to territory in this region.[41] This would ultimately pave the way for 'the ultimate objective of making astronautics a truly world enterprise'.[42] Here, the enduring idea of a globally-connected outer space project was presented against a background of historic British achievement in global exploration:

> Surely it is unthinkable that Britons will not participate in this new exploration or are the New Elizabethans much inferior to the old? Drake, Raleigh, Hudson, Cook, Park, Franklin, Eyre, Burke and Wills, Burton, Baker, Speke, Grant, Darwin, Livingstone, Stanley, Shackleton and Scott – do these names mean nothing anymore? Are Hunt and Hillary to be the last of the line?[43]

In naming a list of pioneering explorers starting with 'old Elizabethans' Drake and Raleigh and ending with Hunt and Hillary, the leaders of the 1953 Mount Everest expedition, Thompson calls upon a lineage of British imperialism in a similar way to which the European Space Agency drew inspiration from their own genealogy of Classical and Enlightenment thinkers half a century later, helping to legitimise space exploration by association with earlier icons of Western civilization.[44] As well as this prevailing imperialist sensibility, Thompson's list alludes to the small but significant cultural movement of New Elizabethanism, which invoked the reign of Elizabeth I as a 'golden age' to 'inspire a similar renaissance in the twentieth century' under Elizabeth II.[45] As such, the ascent of Mount Everest is presented as an iconic British achievement connected to ideas of modernity, monarchy and internationalism that were key traits of New Elizabethanism, in a potent agglomeration of identities that became a defining characteristic of British spaceflight discourse at this time.[46]

The jewel in the crown of the proposed British Commonwealth space project was the Woomera rocket range in South Australia. Described as 'a most impressive asset to Commonwealth research', the desert facility was used initially as a weapons testing ground by the British government in the late 1940s, and during the 1950s also witnessed a series of nuclear weapons tests in the vicinity.[47] Activity at these proving grounds contributed to what has been called 'British defence futurism' in this period of post-war re-armament amid new Cold War tensions that extended into the territories of the British Commonwealth.[48] One advantage of Woomera was that its fallout range lay 'across empty wilderness to the shores of the Indian Ocean' and beyond, a set of conditions that could not be matched in the home territory of Britain.[49] In this way, Woomera was seen as an expansive reserve of otherwise unused land, enrolled into British plans for space exploration through association with a new, modern Commonwealth of Nations. Part of this claim that the Woomera range productively put to use empty land was the necessary discounting of Aboriginal occupation, a legacy of a broader process of agricultural enclosure and 'improvement of the colony' in Australia from the nineteenth century onwards.[50]

While the perceived geographical advantages of using Woomera as a launch-pad to space were never fully put to use, the imaginative landscape of 'Spaceport Woomera' captured the attention of popular and science-fiction writers in the 1950s and 1960s, as 'a remote and exotic location where intrigue, adventure and the inspiration of spaceflight might be found'.[51] One example of Woomera being presented in fiction as a future spaceport was Arthur C Clarke's first science-fiction novel, *Prelude to Space*, which preceded a substantial proliferation of comics, novels, radio plays and television programmes that made use of this scenario. The notion that Clarke's outer-space fiction was anchored to more familiar Earthly geographies of adventure and exploration has been established in Chapter 4 and with *Prelude to Space* being set entirely on Earth in anticipation of the world's first manned space launch, it is even more overtly connected to such discourses. Indeed, this novel was typical of Clarke's early narratives of 'optimistic scientific propaganda' and was dedicated to his 'friends in the BIS', who seem to have inspired some of the fictional characters.[52] The first half establishes London as the administrative centre of an international spaceflight community and repeats the mantra of New Elizabethanism by invoking 'the line that stretched back to Drake and Raleigh' in its exhortation of British spaceflight.[53] The narrative then leads on to a future version of Woomera, where the first manned space launch is taking place:

> Luna City was built by the British government around 1950 as a rocket research base. Originally it had an aborigine name – something to do with spears or arrows.[54]

Here Clarke alludes to the meaning of 'Woomera', the accepted English term for an Aboriginal spear-throwing device, but dismisses this association as an irrelevant myth, preferring instead the Latin derivative 'Luna City'.[55] While indigenous cultures are overwritten, Commonwealth symbolism abounds within the novel, as Clarke places British icons such as the Union Flag and a letter from 10 Downing Street in the heart of the Australian desert. Moreover, in the final pages, the sound of Big Ben chiming out through loudspeakers is described as the space ship 'Prometheus' is finally launched. These somewhat clichéd portrayals of 'Britishness' help to advance this conception of an interplanetary project that is at the same time British and international, towards a framework that exploits the British Commonwealth as the backdrop for interplanetary scientific internationalism. De Witt Douglas Kilgore further notes the significance of this passage in his account of *Astrofuturism*, noting that, in Clarke's foregrounding of these racial and nationalist tropes, 'space itself becomes a frontier with Greenwich and Westminster at its heart'.[56]

One non-fiction publication, *Rockets in the Desert* by children's writer Ivan Southall, presents Woomera in a similar way, using the framework of a British-led Commonwealth space project. 'Woomera', states Southall, who visited the facility by special permission in the early 1960s, 'began in 1945 in England' and was now 'the most advanced space research station of its type on earth'.[57] Southall encourages his young readers to 'join in our great adventure of exploring the heavens', presenting this kind of work as wholesome, healthy and energetic, but also highly dangerous.[58] This presentation of Woomera as a site of Western modernity and adventure sits uncomfortably alongside the fact that Aboriginal lands formed part of the active testing range. Although Southall's book somewhat glosses over such aspects, a more comprehensive account by historian Peter Morton some years later explains the controversy surrounding the construction of the rocket range, whose central line was mapped 'slicing through the Central Aboriginal Reserves' in 1946 (Fig. 5.3).[59]

In these narrations by Clarke, Southall and others, the representation of Woomera as a site of adventure and exploration dominates the significant narratives of resistance and protest that occurred there in the late 1940s. That such narratives took place in the modern context of Cold War rocketry reminds us that geographies of adventure and exploration form an enduring part of contemporary geopolitical discourses and were not just limited to the more traditional Victorian-era colonial narratives.[60] Progress in British

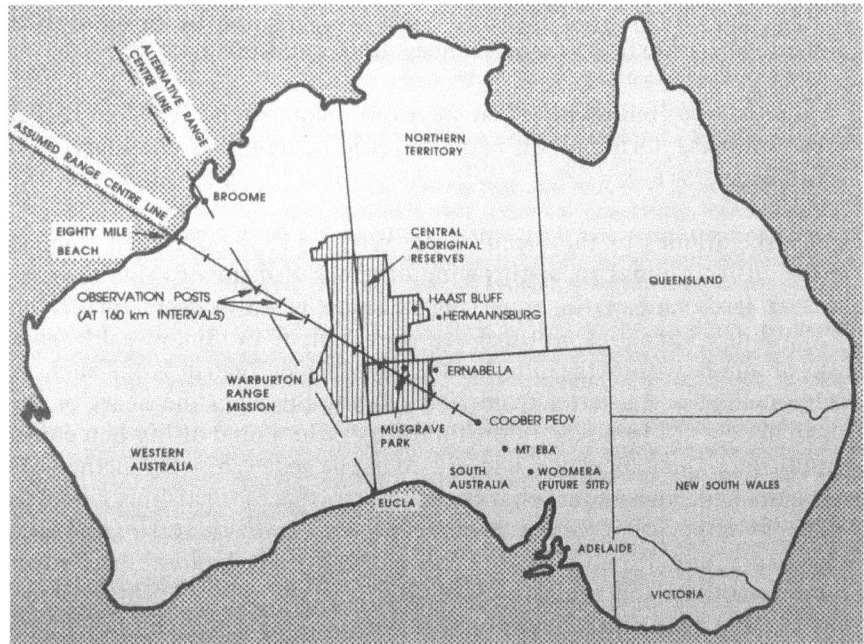

Figure 5.3 Map of Australia showing the Woomera Rocket Range

Source Credit: Commonwealth of Australia Department of Defence

space exploration was not just limited to fiction, however, as scientific and technological research into rocketry became simultaneously caught up in further geopolitical discourse during this period.

Geopolitical pivots: Americanism and Europeanism in British outer space

Recent analysis of the post-war period in outer-space culture has suggested that, while the early-post-war years witnessed an explosion in cultural expressions of outer space in Britain, through media such as comics, television and radio, in later years 'the imaginary that had been building [became] limited by the realities of Sputnik and what followed into space'.[61] If such a shift from an imaginative period of outer-space culture to a technocratic period did occur, it was certainly marked by the spectacular early achievements of the Space Age, with public audiences having been primed on the adventure of outer space actually able to witness television broadcasts of men on the Moon. Such achievements triggered urgent calls for the UK to become involved in outer space, with one *Spaceflight* article in 1960 stating that 'British entry into space activities [...] has now become essential if we are to maintain our position in world technology'.[62] However, while the

UK government gradually became more interested in outer-space projects, the reality was that Britain's global status as a major economic power was receding and its approach to space exploration was largely contingent on a shifting set of international agreements. In this way, the geopolitics of outer space were indicative of Britain's changing place in world affairs. Here, a sense of geopolitical decline was largely established by, or reflected in, the Suez crisis of 1956, where Britain's global influence was effectively downgraded and replaced by that of the United States.[63] For the remainder of this section, two case studies illustrate these themes, each connecting with broader geopolitical discourses relating to, first, the American sphere of influence and second, that of continental Europe.

Ariel 1 and Starfish Prime

Following the success of the early satellite programmes of the Soviet Union and the United States, the UK government, while not committing to a full national space programme, was willing to support a limited agenda of space exploration in collaboration with other allied nations. A first step was the establishment of the Skylark sounding-rocket programme in 1957 as part of the International Geophysical Year. This involved launching experimental rockets into the upper atmosphere from Woomera to measure some of the qualities of the little-known region on the borders of space, including the ionosphere and Van Allen radiation belts.[64] Crucially, research in this area would also support the broader endeavour of ballistic missile development, which was becoming central to global defence agendas as the Cold War intensified. After the establishment of NASA in 1958, the United States was keen to encourage allied nations to establish their own space programmes, collaborating where possible. This new period of co-operation, formalised in treaties such as the 1958 US–UK Mutual Defence Agreement, helped to replace the prior atmosphere of secrecy that had surrounded US programmes in military science, particularly in atomic power and ballistic missile development. Following these initiatives, the Ariel satellite programme was conducted between 1962 and 1979, involving the launch of six satellites through British-American co-operation.[65]

Ariel 1 was successfully launched into orbit on the 26th April 1962 from Cape Canaveral on an American Thor-Delta rocket, carrying out a series of experiments on the upper atmosphere. Over and above its scientific significance, the initial success of Ariel lay in the spectacle and prestige of being Britain's first satellite, as was similarly the case with other 'big science' projects of the post-war period such as Jodrell Bank radio telescope.[66] Indeed, a special publicity committee was established, involving the Foreign Office and Ministry of Science, which promoted Ariel in a series of television and radio programmes.[67] The launch was reported in *The Times*, whose science correspondent noted 'strong signals from Ariel' as it passed over Britain twice in one day.[68] It was referred to in the press as both 'the Anglo-American satellite' and 'the first British satellite', as while the instruments in its payload were designed by a British team, the satellite itself and the launch equipment were all American.[69] The Ariel programme was led by a team of researchers and engineers from British universities, alongside the UK Science Research Council and NASA's Goddard Spaceflight Center. Matthew Godwin has explained how the British authorities, as part of efforts to maintain a place at the 'top table' of international affairs, encouraged university researchers to pursue such studies on the near regions of outer space.[70] This strategy seemed to be paying off as Ariel 1 was able to return data to monitoring stations around the world during its first few months of operation and Britain was thereby able to claim a status as a leading nation in outer-space technology. In this way, Ariel 1 took on a hybrid geopolitical agency, both as the first British satellite, but also as the world's first international satellite. However, as Godwin has noted, while the UK benefited from the prestige of claiming a stake in a joint space project, 'in practice the nature of co-operation was controlled by the Americans', who were in charge of the launch, the rocket and the satellite itself.[71]

Ariel 1 was a short-lived symbol of international collaboration and national prestige in space science, as on the 9th July 1962, the satellite was critically damaged by the US 'Starfish Prime' nuclear test. Through the overarching Project Fishbowl, the United States co-ordinated five hydrogen-bomb detonations up to altitudes of 250 miles over the Pacific Ocean, tests which occurred before the introduction of limited and comprehensive nuclear test ban treaties and the 1967 Outer Space Treaty. Nuclear tests at this altitude, on the borders of space, released charged particles and radiation into the upper atmosphere, which then circulated around the Earth. In Britain, reaction to these tests included a series of strong interventions by Bernard Lovell, the director of Jodrell Bank. While not directly anticipating the effect on Ariel 1, Lovell commented that 'scientists [...] will be filled with dismay at the American proposal to perform a nuclear explosion in [this] region of space', given its likely interference with upper-atmosphere radiation belts, while criticising the 'presumption of moral right to interfere with the environment of the Earth' in this way.[72] While news of Ariel's malfunction emerged in the days and weeks following the incident, the true cause of the damage was not relayed to the British authorities until a month afterwards and was 'initially denied by the US authorities'.[73] An article in *Spaceflight* by NASA's Director of International Programmes the next year somewhat awkwardly glossed over the incident, noting that 'Ariel is providing the first information combining certain aspects of solar radiation and events in the ionosphere', while pointing towards the enhanced role British scientists would take in the subsequent Ariel satellites.[74]

Given the remoteness of these events to typical observers in Britain, the geopolitical discourse of Ariel 1 and its fate at the hands of the Starfish Prime nuclear test was mediated by government press releases and independent journalism. In any case, this episode encapsulates a sense in which there was a conflicting desire for the UK to remain as a leading power in outer space, but also having to negotiate and still be in line with the USA in its space projects. More broadly, it also illustrates how British hopes for a national or internationally-co-operative entry into outer-space activity became entirely conditional on the geopolitical atmosphere of the Cold War. All this notwithstanding, the Ariel programme was undoubtedly important in the development of British space science, with its legacy feeding into later expertise in the British satellite industry, activity which is picked up in Chapters 7 and 8.

ELDO and Blue Streak

At the same time that the US and UK were initiating collaborative work in space science projects, co-operation between the UK and other European countries in space was also seen as worthwhile, with commentators in the BIS claiming that 'Europe would be able to compete on equal terms with the USA and USSR' and highlighting 'a number of such co-operative projects in nuclear science and engineering', which could form a model for European collaboration on spaceflight.[75] This kind of multilateral co-operation led to the formation of two organisations in 1964; the European Launcher Development Organisation (ELDO), which was initially funded primarily by the UK government; and the European Space Research Organisation (ESRO), which eventually merged to form the European Space Agency (ESA) in 1975.[76] The role of ELDO was to develop a European satellite launcher and the resultant hybrid rocket, named 'Europa', consisted of multiple stages, each designed by teams in different European countries. So, while the primary booster rocket was developed from the UK's 'Blue Streak' missile, the second stage 'Coralie' was designed by the French, the third stage 'Astris' was German in design and satellite payloads were designed by the Italians. In total eleven launch tests of Europa were conducted, the first ten at Woomera and the final test at Kourou in French Guiana in 1971. This shift in launch sites came about due to Britain's withdrawal from the project, as it sought instead to develop its own rival satellite launcher, Black Arrow.[77]

When faced with a possible choice of American or European co-operation in space, many of those in the nascent British space industry favoured allegiance to a European project. Val Cleaver, one of the designers of Blue Streak and a central figure in British rocketry in this period, noted 'the disadvantages of needing to thumb a lift' on American space rockets, further suggesting that 'the absence of a European space programme would probably be the beginning of a downhill slide for our science, technology and industry'.[78] Indeed, in later correspondence Cleaver explained that:

> In this space crisis, over the past six years, I have definitely thought and behaved as a European first and an Englishman second.[79]

The BIS continued to support European collaboration throughout the 1960s, with a set of government recommendations submitted in 1965, calling for the UK to 'actively support the development of an integrated Western European Space Programme [which] will eventually make collaborative European–US projects desirable and advantageous'.[80] The acceptance that a move towards Europe would be beneficial for a British space programme reflected a more general trend in the 1960s of European co-operation and pro-Euopeanism in Britain that surrounded the formation of the European Economic Community in 1957 and the UK's eventual accession to the group in 1973. John Krige has argued that the development of the Europa launcher was propelled for political rather than technical reasons, stating that it 'only managed to take shape because of the very specific political situation prevailing in Europe at the time', as the preparations for Britain's entry into the European single market took place.[81] Aware of this geopolitical situation, BIS members in the 1960s expressed a certain satisfaction with British contributions to ELDO and ESRO, but ultimately believed that Britain should be doing more:

> Slowly, almost imperceptibly, Britain is becoming involved in astronautics. Our aims are not coextensive with ELDO's but far more extensive than those of ELDO and ESRO put together [...] Co-operation plus independent action equals maximum progress.[82]

Indeed, towards the end of the 1960s, the Europa programme came to be seen as 'confused, unplanned and ineffectual' by the BIS, who were cognisant of the fact that the rocket had failed to put a fully-functional satellite into orbit.[83] Val Cleaver had referred to the British space situation in a letter to Arthur C Clarke as 'a tragi-comedy', citing fears that France would soon overtake Britain as 'the leading European technological nation', as well as 'doubts, delays and bureaucratic fumbles' on the part of the UK government.[84] In the end, although it established some of the groundwork for the later, more successful *Ariane* rockets, the Europa project was seen both as a political and a technical failure and funding was cancelled in the run-up to the final testing in 1971. It was, however, a remarkable symbol of international co-operation, a physical manifestation of the broader geopolitical project of European integration.

Britain's response to leaving ELDO came in the form of the Black Arrow, which did succeed in placing the experimental satellite 'Prospero' into orbit in 1971. Like Europa, however, Black Arrow also had its funding withdrawn in the lead-up to this final launch, due to its unreliable performance record, and Prospero remains the only satellite Britain has built and launched in the Space Age, in some ways marking an end-point to a British space programme that began with Ariel 1 a decade earlier.[85] After this, it became apparent that a more limited involvement in outer space was what was left for Britain, despite the substantial weight of optimism and international goodwill that had characterised what Arthur C Clarke called 'the heroic period of the space age [that] lay between 1935 and 1955'.[86] Having sponsored European efforts to develop a satellite launcher system and through its alliance with the American space programme, we can see how, in this period, British outer-space initiatives were highly contingent on a range of geopolitical associations. This reflected and in many ways came to symbolise Britain's sometimes uneasy international identity as Atlantic partners with the United States, Continental partners with European nations and Commonwealth partners with newly independent states of the former British Empire.

Conclusion: British astropolitics and the limits of modernity

Was there such a thing as a British space programme? This chapter certainly has identified a successive development of concepts, partnerships and technologies that led to the successful launch of a satellite on a British space rocket. However, such a simplistic story would not take account of the multitude of failures, anxieties and contradictions that largely characterised such a programme and its complex geopolitical contingencies. Moreover, this chapter has explored the ways in which British outer-space activities in the post-war period became connected to international networks of science, technology and politics, demonstrating that an understanding of the geopolitics of outer space should be seen as intertwined with historically and culturally specific contexts and emphasising the effectiveness of critical astropolitics in explaining the relevance of the British space programme in this period. Therefore, in order to fully understand how and why British ambitions in space reached their limit with a series of semi-failed collaborative projects in the 1960s and early 1970s, it has been necessary to explore the social, cultural and political aspects of this broadly-conceived programme. This has involved drawing from a diverse array of narratives, representations and performances, from the stage-managed formation of the International Astronautical Federation, to the cultural representation of Woomera as a site of danger and adventure in outer space, and the political management of nuclear fallout in the pre-treaty era of the Cold War. In doing this, this chapter has revealed how the implied utopianism of internationalism in early-post-war British cultures of outer space ultimately became limited by nationalist, colonialist and geopolitical concerns in later years, illustrating the fine balance between 'the abstractions of internationalism and the geographical and historical specificities of its performance'.[87] In terms of the British Interplanetary Society, its members tapped into various discourses of national and international identity formation, illustrating and performing cultural tropes such as New Elizabethanism, utopian internationalism and conservative pro-Europeanism. It thereby reveals itself as an organisation that was not just parochial in scope, clinging to insular ideas of British leadership in space, but did expose itself to broader horizons, even as these visions became tempered at times with nostalgic references to Empire. In the post-Apollo era in outer space, many observers have noted how enthusiasm for space seemed to wane.[88] The next chapter explores the circumstances behind a yet more ambitious plan designed by members of the BIS, this time looking beyond the Solar System entirely.

Notes

1 William Breuer, *Race to the Moon – America's Duel with the Soviets* (Westport: Praeger, 1993).
2 *Militarizing Outer Space: Astroculture, Dystopia and the Cold War*, ed. by Alexander Geppert, Tilmann Seiebeneichner and Daniel Brandau (Palgrave Macmillan, 2020); Roger D Launius, 'What Are Turning Points in History, and What Were They for the Space Age?', in *Societal Impact of Spaceflight*, ed. by Roger D Launius and Steven J Dick (Washington DC: NASA, 2007), 19–40.
3 Everett C Dolman, *Astropolitik: Classical geopolitics in the space age* (London: Frank Cass, 2002), p.168.
4 Randall R Correll, and Simon P Worden, 'The Demise of US Spacepower: Not with a Bang but a Whimper', *Astropolitics: The International Journal of Space Politics & Policy*, 3 (2005), 233–264; John J Klein, 'Space Warfare: A Maritime-Inspired Space Strategy', *Astropolitics: The International Journal of Space Politics & Policy*, 2 (2004), 33–61; Martin E B France, 'Back to the Future: Space power theory and A T Mahan', *Space Policy*, 16 (2000), 237–241.
5 Sheng-Chih Wang, 'The Making of New "Space": Cases of Transatlantic Astropolitics', *Geopolitics,* 14 (2009), 433–461.
6 Joan Johnson-Freese, *Heavenly Ambitions: America's Quest to Dominate Space* (Philadelphia: University of Philadelphia Press, 2009); *Securing Outer Space*, ed. by Natalie Bormann and Michael Sheehan (London: Routledge, 2009); Alison Williams, 'Beyond the Sovereign Realm: The Geopolitics and Power Relations in and of Outer Space', *Geopolitics,* 15 (2010), 785–793.
7 Fraser MacDonald, 'Anti-Astropolitik: Outer Space and the Orbit of Geography', *Progress in Human Geography,* 31 (2007), 592–615 (p.608); Jonathan Havercroft and Raymond Duvall, 'Critical astropolitics: the geopolitics of space control and the transformation of state sovereignty', in *Securing Outer Space*, ed. by Natalie Bormann and Michael Sheehan (New York: Routledge, 2009), 42–58; Nayef Al-Rodham, 'The Meta-Geopolitics of Outer Space', in *The Palgrave Handbook of Society, Culture and Outer Space*, ed. by Peter Dickens and James S Ormrod (Basingstoke: Palgrave Macmillan, 2016), 123–166.
8 MacDonald, 'Anti-Astropolitik'.
9 John Agnew and Gearóid Ó Tuathail, 'Geopolitics and discourse – Practical geopolitical reasoning in American foreign policy', *Political Geography,* 11 (1992), 190–204 (p.192).
10 Joanne Sharp, 'Hegemony, Popular Culture and Geopolitics: The Reader's Digest and the Construction of Danger', *Political Geography,* 15 (1996), 557–570 (p.557).
11 Daniel Sage, *How Outer Space Made America: Geography, Organization and the Cosmic Sublime* (Farnham: Ashgate, 2014).
12 Daniel Sage, 'Framing Space: A Popular Geopolitics of American Manifest Destiny in Outer Space', *Geopolitics,* 13 (2008), 27–53.
13 Jason Dittmer, 'Colonialism and Place Creation in "Mars Pathfinder" Media Coverage', *The Geographical Review,* 97 (2007), 112–30 (p.112).

14 Lane K Maria D Lane, 'Geographers of Mars – Cartographic Inscription and Exploration Narrative in Late Victorian Representations of the Red Planet', *Isis,* 96 (2005), 477–506; Shaun Huston, 'Murray Bookchin on Mars! The Production of Nature in Kim Stanley Robinson's Mars Trilogy', in *Lost in Space – Geographies of Science Fiction,* ed. by Rob Kitchin and James Kneale (London: Continuum, 2002), 167–179.

15 Peter Redfield, *Space in the Tropics – from Convicts to Rockets in French Guiana* (Berkeley: University of California Press, 2000), p.148.

16 Redfield, *Space in the Tropics* (p.127).

17 Redfield, *Space in the Tropics* (p.139).

18 Peter Dickens, 'The cosmos as capitalism's outside', in *Space Travel and Culture: From Apollo to Space Tourism,* ed. by M Parker and D Bell (Oxford: Blackwell, 2009), 66-82 (p.79).

19 Peter Redfield, 'The Half-Life of Empire in Outer Space', *Social Studies of Science,* 32 (2002), 791–825.

20 Arthur Valentine Cleaver, 'The Interplanetary Project', *Journal of the British Interplanetary Society,* 7, 1 (1948), 21–37.

21 Bob Parkinson, *Interplanetary – a History of the British Interplanetary Society* (London: British Interplanetary Society, 2008), p.38 [Thompson joined the BIS in 1946, and was one of the Society's most active members, acting as President (1979–1982) and editor of *JBIS* (1957–1965). With a background in chemical engineering, he is notable for his translations of Russian technical literature. Shepherd joined the BIS in 1935 and was part of the first BIS delegation to the IAF in 1950, later becoming President of the IAF (1956–7)].

22 International Astronautical Federation, 'About the International Astronautical Federation', *International Astronautical Federation* (2020) http://www.iafastro.org/about/ [2ⁿᵈ October 2016].

23 'First International Astronautical Congress, Paris, 1950', *Journal of the British Interplanetary Society,* 10 (1951), 1–4 (p.1).

24 Leslie Shepherd, 'Prelude and First Decade, 1951–1961', *Acta Astronautica,* 32 (1994), 475–499 (p.478).

25 'First International Astronautical Congress', (p.4).

26 Fraser MacDonald, *Escape from Earth – a Secret History of the Space Rocket* (London: Profile, 2019).

27 Leslie Shepherd, 'The International Astronautical Federation', *Spaceflight,* 1 (1957) 159–153 (p.160).

28 Alastair Pearson and others, 'Cartographic Ideals and Geopolitical Realities: International Maps of the World from the 1890s to the Present', *The Canadian Geographer,* 50 (2006), 149–76 (p.162).

29 Guenter Loeser, 'The First Task of an International Federation – An Institute for Astronautics', *Journal of the British Interplanetary Society,* 10 (1951) 156–148 (p.146).

30 Loeser, 'The First Task of an International Federation' (p.146).

31 'Second International Astronautical Congress, London, 1951', *Journal of the British Interplanetary Society,* 10 (1951), 318–330 (p.318).

32 'Second International Astronautical Congress' (p.318).

33 Ruth Craggs and Martin Mahony, 'The Geographies of the Conference', *Geography Compass,* 8 (2014), 414–430 (p.414); 'Second International Astronautical Congress' (p.326).

34 'Second International Astronautical Congress' (p.326).

35 Becky Conekin, *'The Autobiography of a Nation' – the 1951 Festival of Britain* (Manchester: Manchester University Press, 2003), p.57.

36 Denis Cosgrove, *Apollo's Eye: A Cartographic Genealogy of the Earth in the Western Imagination* (Baltimore: John Hopkins University Press, 2001).

37 *Dada Culture – Critical Texts on the Avant-Garde*, ed. by Dafydd Jones (New York: Rodopi).

38 Alison Tucker, *From Imagination to Reality – An Audio History of the British Interplanetary Society* (Milton Keynes: Delta Vee Media, 2008) [on CD].

39 E D G Andrews, 'British Participation in Space Research', *Spaceflight*, 3 (1961) 57–67 (p. 59).

40 Gordon V E Thompson, 'The British Space Research Programme', *Journal of the British Interplanetary Society* 17 (1959), 41 (p.41).

41 Kenneth Gatland, 'Towards a Commonwealth Space Agency', *Spaceflight* (1959), 35–38 (p.37); Klaus Dodds, 'Assault on the Unknown: Geopolitics, Antarctic Science and the International Geophysical Year (1957–8)', in *New Spaces of Exploration – Geographies of Discovery in the Twentieth Century*, ed. by Simon Naylor and James Ryan (London: I B Tauris, 2010), 148–172.

42 Gatland, 'Towards a Commonwealth Space Agency' (p.35–6).

43 Gordon V E Thompson, 'Britain and Space', *Journal of the British Interplanetary Society*, 17 (1959) 157–158 (p.158).

44 Redfield, 'The Half-Life of Empire'.

45 Richard Weight, *Patriots – National Identity in Britain 1940–2000* (London: Macmillan, 2002), p.226.

46 Peter Hansen, 'Coronation Everest: The Empire and Commonwealth in the "Second Elizabethan Age"', in *British Culture and the End of Empire*, ed. by Stuart Ward (Manchester: Manchester University Press, 2001), 57–72.

47 Arthur Valentine Cleaver, 'Woomera', *Spaceflight*, 1 (1957) 103–106 (p.106); Elizabeth Tynan, *Atomic Thunder: The Maralinga Story* (Sydney: University of New South Wales Press, 2016).

48 David Edgerton, *Warfare State: Britain, 1920–1970* (Cambridge: Cambridge University Press, 2006), p.232.

49 Len J Carter, 'Anglo-Australian Long Range Weapon Project', *Journal of the British Interplanetary Society*, 9 (1950), 1–5 (p.1).

50 Chris Gibson, 'Cartographies of the Colonial/Capitalist State: A Geopolitics of Indigenous Self-Determination in Australia', *Antipode,* 31 (1999), 45–79 (p.60); Alice Gorman, '*La terre et l'espace*: rockets, prisons, protests and heritage in Australia and French Guiana', *Archaeologies: Journal of the World Archaeological Congress* 3 (2007), 153–168.

51 Kerrie Dougherty, 'Spaceport Woomera: The Anglo-Australian Vision of Woomera Rocket Range', *Quest – The History of Spaceflight,* 22 (2008), 1–19 (p.14).

52 Peter Nicholls, 'Clarke, Arthur C', in *The Encyclopedia of Science Fiction*, ed. by Peter Nicholls and John Clute (London: Granada, 1981), 121–22 (p.121); Neil McAleer, *Odyssey: The Authorised Biography of Arthur C Clarke* (London: Victor Gollancz 1992), p.70.

53 Arthur C Clarke, *Prelude to Space* (London: Sidgwick and Jackson Ltd., 1953), p.11.

54 Clarke, *Prelude to Space* (p.104).

55 Dougherty, 'Spaceport Woomera'.

56 De Witt Douglas Kilgore, *Astrofuturism: Science, Race and Visions of Utopia in Space* (Philadelphia: University of Pennsylvania Press, 2003), p.122.

57 Ivan Southall, *Rockets in the Desert – the Story of Woomera* (Sydney, Australia: Angus and Robertson, 1965), p.23 [An earlier version of this book, *Woomera* (Sydney, Angus and Robertson) 1962, was intended for adult readership].

58 Southall, *Rockets in the Desert* (p.78).

59 Peter Morton, *Fire across the Desert – Woomera and the Anglo-Australian Joint Project, 1946–1980* (AGPS: Canberra, 1989), p.71.

60 *New Spaces of Exploration – Geographies of Discovery in the Twentieth Century*, ed. by Simon Naylor and James Ryan (London: I B Tauris, 2010); Richard Phillips, *Mapping Men and Empire: A Geography of Adventure* (London: Routledge, 1997).

61 Doug Millard, 'A Grounding in Space: Were the 1970s a Period of Transition in Britain's Exploration of Outer Space?', in *Limiting Outer Space: Astroculture after Apollo*, ed. by A Geppert (London: Palgrave Macmillan, 2018), 79–99 (p.94).

62 'Shaping the Space Programme' *Spaceflight*, 2 (1960), 239–246 (p.239).

63 Weight, *Patriots*.

64 Matthew Godwin, *The Skylark Rocket – British Space Science and the European Space Research Organisation 1957–1972* (Paris: Beauchesne, 2007).

65 Harrie Massie and M O Robins *History of British Space Science* (Cambridge: Cambridge University Press, 1986).

66 Jon Agar, *Science and Spectacle – the Work of Jodrell Bank in Post-War British Culture* (Amsterdam: Harwood Academic Publishers, 1998).

67 Matthew Godwin, 'Britnik: How America Made and Destroyed Britain's First Satellite', in *New Spaces of Exploration – Geographies of Discovery in the Twentieth Century*, ed. by S Naylor and J Ryan (London: I B Tauris, 2010), 173–195.

68 'Strong Signals from Ariel', *The Times*, 28th April 1962, 55377, p.8.

69 'Bristol Make Further Space Equipment', *The Times*, 29th May 29 1962, 55403, p.6.

70 Godwin, *The Skylark Rocket*.

71 Godwin, 'Britnik' (p.177).

72 'Dismay over Proposed Explosion in Space', *The Times*, 2nd May 1962, p.12.

73 'UK 1 Acts Up, Breaks Down', *Missiles and Rockets*, 30th July 1962, p.13; Godwin, 'Britnik' (p.187).

74 Arnold W Frutkin, 'Progress in International Co-operation in Space Research', *Spaceflight*, 5 (1963), p.207.

75 'Shaping the Space Programme', (p.244); 'Participation of the UK in Space Flight Development – European Collaboration', *Journal of the British Interplanetary Society*, 17 (1960) 237 (p.237).

76 ELDO was funded by the UK (£27m), France (£17m), Germany (£15.5m), Italy (£7m), Belgium (£2m), the Netherlands (£2m), and Australia providing the rocket range and supporting facilities. *Spaceflight*, 5 (1963), p.8.

77 John Krige and Arturo Russo, *A History of the European Space Agency, 1958–1987, Volume 1 – The Story of ESRO and ELDO, 1958–1973* (Noordwijk: ESA: 2000).

78 Arthur Valentine Cleaver, 'The Implications of Not Having a Space Programme', *Spaceflight,* 4 (1962), 22–24.

79 Andrew Chatwin, 'Finale and Reprise, 1974–1977', in *Val Cleaver (1917–1977) – a Very English Rocketeer*, ed. by Andrew Chatwin (London: British Interplanetary Society, 2014), 124–145 (p.137).

80 'Recommendations by the Council of the BIS to HM Government', *Spaceflight*, 7(1965), 111–117 (p.111).

81 Krige and Russo, *A History of the European Space Agency* (p.80).

82 Gordon V E Thompson, 'The Future of British Astronautics', *Journal of the British Interplanetary Society* 18 (1962), 317–318 (p.317).

83 'A Space Policy for Britain', *Spaceflight*, 10 (1968) 56–57 (p.56).

84 Chantilly, Virginia, Steven F Udvar-Hazy Center, National Air and Space Museum, Smithsonian Institution, Arthur C Clarke Collection of Sri Lanka, Correspondence Box 2 and 3, Val Cleaver to Arthur C Clarke, 7th June 1963 and 13th May 1962.

85 Millard, 'A Grounding in Space'.
86 Arthur C Clarke, *Voices from the Sky: Previews of the Coming Space Age* (London: Victor Gollancz, 1966), p.167.
87 Jake Hodder, Stephen Legg, and Michael Heffernan, 'Introduction: Historical Geographies of Internationalism, 1900–1950', *Political Geography,* 49 (2015), 1–6 (p.1).
88 *Limiting Outer Space: Astroculture after Apollo,* ed. by Alexander Geppert (London: Palgrave Macmillan, 2018); Roger D Launius, *Apollo's Legacy – Perspectives on the Moon Landings* (Washington DC: Smithsonian Books, 2019).

6   Interstellar exploration

Project Daedalus and the extra-solar universe

Following the limited achievements of the British satellite launcher project, a return to expansive speculative theorising was perhaps inevitable among the UK spaceflight community. Indeed, by the 1970s, as the Apollo Moon landings took place to global acclaim, and probes such as Mariner 2 and Pioneer 10 had started to explore the inner and outer planets of the Solar System, the realms of possibility in space exploration seemed to be expanding beyond the limits of Earth's orbit. At this time, members of the British Interplanetary Society decided to make a case for interstellar travel, aiming to produce a design concept for a spacecraft that would be capable of reaching one of the nearest stars in the galaxy within a period of fifty years, seen as the professional life-span of a typical scientific researcher. Under the name of Daedalus, the legendary craftsman of ancient Greece, a starship was designed that would use 'present-day technology and *reasonable* extrapolation for near-future capabilities'.[1] The vehicle itself would be resourced from, and constructed in, the Solar System environment, requiring a new synthesis in the understanding of our planetary neighbourhood. The destination of Barnard's Star, the fourth-closest star to the Sun, was decided upon, with a nuclear fusion pulse rocket accelerating Daedalus to twelve per cent of the speed of light. Daedalus was to be uncrewed, returning data on Barnard's Star and the interstellar environment back to Earth, before its trajectory would propel it further out into interstellar space.

Project Daedalus attracted significant interest from the wider scientific community, and remains one of the most comprehensive examples of the conceptualisation of spaces outside of the Solar System since the advent of the Space Age, while it has acquired iconic status within the BIS. Throughout this chapter, it is argued that thinking through notions of place, exploration and environment provide windows on the conceptualisation of interplanetary and interstellar space, in building on this book's underlying contention that geographical concepts are central to understandings of outer space. As such, attention is drawn to an emerging cross-disciplinary body of research that has critically examined the cultures and politics of outer space with an awareness of the significance of place, including in anthropology, science and technology studies, and environmental history.[2] The chapter first outlines the broader societal and scientific contexts in which interstellar exploration emerged as a viable concept, while also tracing the development of the idea from the immediate post-war years through to the mid-1970s. Here, concepts of exploration were expanded to anticipate the search for a possible home for humanity beyond the Solar System, and the potential existence of alien life in the Universe. The chapter then turns to certain aspects of Project Daedalus itself, specifically its destination, its understanding of the Solar System and its reception among the scientific community. This helps to identify how space exploration, both conceptually and in practice, facilitated ways of seeing other worlds as real places and enrolled new geographies of the Solar System environment into expansive future narratives of humankind.

Geopolitics, extra-terrestrial intelligence and the promise of the atom

In understanding the nature of interstellar space, cosmologists have drawn from a rich intellectual history, including concepts that have bridged the gap between science and the imagination. In general terms, since the scientific revolution in the Western world, the Universe has been conceptualised according to principles that have de-centred the Earth, following in the tradition of Galileo, Kepler and Copernicus. Ideas such as the universal effects of gravity, the form and distribution of galaxies, the plurality of worlds, and the notion of relativity have since deepened understandings of the interstellar environment, alongside technological developments in telescopic observation, spaceflight, spectroscopy and data telemetry. Further discoveries, including the existence of black holes, pulsars and other types of star, as well as the theory of an expanding Universe, have all informed discourses of astronomy and cosmology throughout the twentieth century.

Of particular relevance to the concept of interstellar travel has been the problem of extra-terrestrial life. With NASA's two Viking probes landing on Mars in 1976, revealing unprecedented images that suggested a lifeless and barren planet, attention turned to the stars as possible harbingers of life in the Universe. Prior to this, the Italian-American physicist Enrico Fermi (1901–1954) had postulated famously that, if the Universe was so massive and ancient that life was abundant in the cosmos, across a full spectrum of technological and spiritual development, then why hadn't extra-terrestrial visitors already arrived on Earth? This question, reportedly proposed in casual conversation with co-workers in 1950 at Los Alamos nuclear research laboratories in the United States, became known as Fermi's Paradox, and would have profound implications for the proposition of interstellar travel.[3] If an interstellar vessel could be designed at a proof-of-concept level, then a reasonable assumption could be made as to the possibility of contact between forms of life across the galaxy and beyond, leaving Fermi's Paradox to be still unresolved. If such a project was deemed impossible, then it might reasonably be concluded that contact between intelligent life forms in the Universe could never occur. In this way, the dual concepts of life in the Universe and interstellar travel became closely intertwined.

To date, only five human-made objects have been sent on trajectories that will take them beyond the boundaries of the Solar System, all of which are space probes designed by NASA's Jet Propulsion Laboratory to study the outer planets and other phenomena at the edges of the Solar System. The furthest of these, Voyager 1, was classified as crossing into interstellar space in 2012, having completed its principal mission of enhancing human knowledge about the outer planets of the Solar System.[4] Probes such as these, therefore, represent the closest humanity has come to directly exploring interstellar space, and have been studied not just for their scientific mission data on planetary bodies and other astrophysical phenomena, but also for their social, cultural and political relevance on Earth. In anticipation of the fact that these objects would leave the Solar System and continue into cosmic space, it was decided that the Pioneer and Voyager probes would carry coded information about human culture and society, in the however unlikely scenario that they would eventually be intercepted by intelligent species from elsewhere in the Universe. Although the implementation of this project is generally credited to the science populariser and Cornell University professor Carl Sagan (1934–1996), the British space advocate Eric Burgess (1920–2005) has claimed to have given Sagan the idea. A leading member of the British Interplanetary Society after the Second World War, in the early 1970s, Burgess was working as a technology reporter for the *Christian Science Monitor*, and had been living in the United States for a number of years. While reporting on the final testing of Pioneer 10 at TRW Systems, California, in November 1971, Burgess recalled:

> I suddenly became aware that this small but intricate machine would actually leave our Solar System [...] I thought of the importance of the spacecraft carrying a special message from mankind [...] that would tell any finder of the spacecraft a million or even a billion years hence that planet Earth had evolved an intelligent species that could think beyond its own time and beyond its own Solar System.[5]

Burgess noted how 'Carl's eyes lit up' as he explained the idea to him for the first time, aware as he was of Sagan's connections in the space science community.[6] Thereby, having successfully lobbied decision-makers at the Jet Propulsion Laboratory, Sagan, in collaboration with Cornell colleagues Frank Drake and Linda Salzman Sagan, designed the 'Pioneer plaque', an engraved aluminium plate that was affixed to Pioneer 10 and 11. These plates display diagrams of the Earth's place in the Solar System, a 'stellar map' of fourteen pulsars visible from Earth, a representation of the elemental make-up of hydrogen, and an image of a man and a woman, drawn to the scale of the probe itself. In this way, after the spark of Eric Burgess' initial idea, interstellar communication became more intimately associated with spaceflight design.

The inscriptions on the Pioneer plaques, Sagan maintained, would act as 'a hopeful symbol of a vigorous society on Earth', while acting as objective, universal indicators of intelligence.[7] However, contrary to this goal, these symbols have since been recognised as subjective, partial and socially embedded indicators of human culture.[8] In particular, the images of the human figures have been criticised as conforming to ethnic, sexual and gender stereotypes, although Sagan and his colleagues had acknowledged some of these limitations. Perhaps mindful of such responses, Sagan curated a more extensive set of messages for the two later Voyager probes, this time in the form of a 'golden record', a gold-plated copper phonograph disk, upon which were inscribed data representing 115 images, music and other sounds, designed to 'portray the diversity of life and culture on Earth'.[9] The records were affixed to the sides of the Voyager spacecraft, along with instruments and instructions for playing them. What these aspects of space probe design tell us is that, as with terrestrial forms of exploration, aspects of culture and politics in human society act as an integral part of scientific endeavour in outer space. Indeed, while the data sent back to Earth from these probes have become invaluable in scientific understandings of the Solar System, their ongoing presence as emissaries of humankind into the stellar realm has continued to convey a cultural resonance that is perhaps even more significant in popular understandings of outer space on Earth.

Well before the design of the Pioneer and Voyager probes, diverse aspects of space exploration had been discussed within the British Interplanetary Society. In the Society's pre-war period, articles were published in the *BIS Journal* on themes such as interplanetary communication, extra-terrestrial life, and spaceship design, showing how the technical features of spaceflight could be considered alongside the more theoretical aspects (see Chapter 3). In 1952, one of the Society's long-standing members, Leslie Shepherd (1918–2012), contributed a paper on 'Interstellar Flight', the first known technical study on this topic. The motivations for thinking beyond the realms of the Solar System for the first time in such detail were, for Shepherd, associated with big ideas such as the plurality of worlds, the possibilities of extra-terrestrial life and the long-term future of humankind:

> There must be many who cannot derive complete spiritual satisfaction from the picture of mankind spending its whole existence in one single infinitesimal planet with no contact with other species who may people countless other worlds of the Universe [...] Humanity dispersed over many worlds appears more secure than humanity crowded on one single planet.[10]

In writing about feelings of 'spiritual satisfaction', Shepherd reveals some of the underlying urges for thinking in terms of the cosmic whole, articulating a sense of secular transcendence in the prospect of interstellar travel. Such themes were picked up a decade later in long-standing BIS member James Strong's book *Flight to the Stars*, which described interstellar flight as 'one of the greatest challenges of Nature', that would ensure the survival of humanity in the long term.[11] This sense of an enlarged realm of existence, alongside a comparative impression of the smallness of the Earth, is reflective of long-established theorisations of the Earth's place in the Universe, connecting notions of the sublime with themes of transcendence and spirituality.

However expansive such themes were, early interventions on the topic of interstellar flight were still grounded firmly in the specific geographical and historical context of their writing. As examined in the previous chapter, the geopolitics of the Cold War were highly significant in the development of both atomic power and rocket propulsion, technologies that were assigned a conflicted identity as both harbingers of high-tech mass-destruction, through the spectre of ballistic nuclear missiles, and as keys to the future of human progress, through opportunities such as atomic power generation and space exploration.[12] The unique conceptual challenges of interstellar travel brought these two technologies together in unexpected ways, with nuclear-powered rockets seen as the only feasible means to reach the necessary speeds for viable interstellar travel. This is because, in comparison to standard hydrocarbon-fuelled rockets, a relatively low mass of fuel is needed to propel the spacecraft over long distances, through a consistent and predictable release of energy.[13] In the United States, concepts of nuclear rocket propulsion were advanced by scientists who had been part of the Manhattan Project during the Second World War. In 1955, Cornelius J Everett and Stanislaw M Ulam of Los Alamos Scientific Laboratory drafted a memorandum outlining a method of propulsion that used nuclear detonations.[14] In this way, a spacecraft with a shock-absorbing 'pusher plate' would be able to ride the successive shock waves resulting from a controlled series of external explosions. Emerging from this concept, from 1958 to 1965, Project Orion was conceived of and directed by the British-American mathematician and physicist Freeman Dyson, receiving significant funding in the post-Sputnik era from the United States Air Force. This was a concept for sending a crewed spaceship to Mars and the outer planets of the Solar System utilising the nuclear bomb propulsion concept. Dyson later commented that his motivations for the scheme lay in 'enlarging the domain of life' throughout the Solar System, with Orion being large enough to convey a colony of settlers to Mars and other worlds.[15] It was eventually cancelled due to the changing political climate that accompanied the signing of the Nuclear Test Ban Treaties in the 1960s, particularly the realisation of the likely dangers of nuclear fallout that would result from the testing and launching of any such vehicle.[16]

While it was conceived in the same atmosphere of opportunity surrounding atomic power in the post-war years, Leslie Shepherd's starship concept for the *BIS Journal* was written before much of the Los Alamos research and the Orion project were completed and declassified.[17] As well as being the first technical study of interstellar travel, it also represents one of the earliest design concepts for space travel by nuclear propulsion. That such plans appeared in the UK should come as little surprise, given that, in the early 1950s, it was one of the leading nations in nuclear research. The UK government had been refining plutonium at Windscale in Cumbria since 1950, while also testing atomic weapons and, later, the hydrogen bomb.[18] Shepherd, who was just beginning his professional career at the Atomic Energy Research Establishment in Harwell, Essex, was well-placed to make the connections between nuclear propulsion and interstellar travel.[19] Deriving power from 'known nuclear reactions', Shepherd calculated that it would be possible to reach the nearest stars in a matter of centuries.[20] As a way of making sense of this extended timeframe, Shepherd used the analogy of 'geological eras', in which 'centuries or millennia are small intervals'.[21] This suggests the influence of Olaf Stapledon, who, in his 1930s science-fictional works, recognised the significance of thinking about space and time together in the cosmic context (see Chapter 2). Indeed, Stapledon had only four years earlier delivered his lecture on 'Interplanetary Man' to the BIS, which included speculations about interstellar travel, noting that 'the shortest of interstellar voyages would certainly take a very long time, in fact, thousands or millions of years'.[22] Accordingly, Shepherd wrote of the need for 'a completely new philosophy of exploration' in order to comprehend such an extended temporal range.[23] Rather than thinking in terms of heroic individual endeavour in space exploration, his proposition involved a 'generation ship', aboard which an entire community of space-faring people would live, die, and pass on their mission to new generations. Such a mission might last up to thirty life-spans, meaning that the majority of the spaceship's inhabitants would live out their entire lives in the course of the epic journey, neither witnessing their home planet of Earth nor any future destination world to be reached. A crucial facet of this generational mission would, according to Shepherd, be 'a very careful control of population', with the inhabitants being 'subjected to a degree of discipline not maintained in any existing community'.[24] Unlike Stapledon, Shepherd did not expand on the moral and ethical consequences of space travel, only hinting at the possibility of a clean slate of morality and ethics in outer space. Nevertheless, his proposals did achieve a synthesis of technical and philosophical questions that was to become a significant feature of later plans for interstellar flight.

Whereas ballistic missile technology and atomic weapon development continued in tandem for much of the Cold War period, the exploration of space, having reached a number of landmark achievements, was ultimately limited in the post-Apollo era to a series of select, specialist programmes such as the space shuttle and the planetary probes.[25] Developing the concept of interstellar travel therefore became reliant on the informal work of enthusiasts such as Shepherd and, later, the Daedalus committee. Indeed, thinking back to Enrico Fermi and his work at Los Alamos raises further questions about the circumstances under which serious speculation about interstellar travel and extra-terrestrial life could occur. At this time, military-industrial research in both the United States and Western Europe was fuelled by Cold War anxieties about arms proliferation and territorial control, including nuclear fission research, ballistic missile technology and space science. By the 1970s, in the UK, members of the BIS who were part of the Daedalus design committee included specialists who had forged careers in the aerospace and military engineering sectors, with companies such as Hawker Siddeley Dynamics and the British Aircraft Corporation, as well as government scientific departments and agencies such as Appleton Laboratory and the Rocket Propulsion Establishment. In this way, expertise in nuclear propulsion, rocketry and aerospace engineering were brought together in a spin-off from various defunct or ongoing British Cold War defence programmes.[26] Indeed, as part of an oral history project on British scientific careers, a key member of the Daedalus team, Dr. Bob Parkinson, indicated that Daedalus was a 'spare time' project, with contributions made voluntarily, and with meetings held in a pub at Aston Clinton in Buckinghamshire during the mid-1970s.[27] This type of activity echoed some of the past traditions of the BIS in Liverpool and London (see Chapters 3 and 4). In such ways, speculative projects concerning interstellar travel and the search for extra-terrestrial intelligence came about in a particular set of circumstances related to geopolitics, national funding of science and engineering and related cultures of work and leisure.

Shepherd's article in 1952 acted as an important precursor to the Daedalus project, and helps to demonstrate how the concept of interstellar travel has implications for a whole range of issues concerning the future of humankind, incorporating biopolitics, human-environment relations, as well as broader social and political issues. A plethora of technical problems were also introduced, such as interstellar navigation, propulsion technologies, and many other unforeseen design issues associated with such an ambitious plan. Having emerged from the specific context of Britain in the immediate post-war period, Shepherd's proposal also resonated with more enduring features of the relationship between humanity and the cosmos, such as the contemplation of life elsewhere in the Universe, and the possibility of humankind destroying itself completely. The paper was initially forgotten in the annals of the British Interplanetary Society, until it was re-visited some twenty years later in the lead-up to Project Daedalus. The following sections will examine aspects of this project, considering the ways in which concepts of place, environment and society became embedded in the study.

Barnard's Star: imagining exo-planetary space

While Daedalus was designed to be an uncrewed vessel, an integral part of its mission was to send data back to Earth from a fly-past of its destination star, and some of the earliest discussions of the project concerned the selection of its destination. Since the aim was to test the concept of inter-stellar flight, the objective had to involve travelling to one of the closest stars in our cosmic neighbourhood. Of these, the red dwarf Barnard's Star was chosen, being the fourth-closest star to the Sun at around six light-years away, after the three stars that comprise the Alpha Centauri system, at around four light-years away. At an open meeting of the BIS in London in January 1973, committee members Alan Bond, Tony Martin, James Strong and Tony Lawton outlined various aspects of the proposed design for Daedalus, including problems concerning propulsion, navigation and radiation, before answering questions from an audience of some 120 peo-ple.[28] Here, Martin explained that Barnard's Star was chosen because of the understanding that it had an attendant planetary system, aware as he was of recent developments in astronomical observation. For some members of the BIS, the prospect of exploring interstellar space carried an additional signif-icance for humanity's far future. In 1970, James Strong postulated that, with an expanding Sun set to make the Earth inhospitable in millennia to come, 'the human race faces extinction unless some part of it can escape to the stars'.[29] In examining options for interstellar navigation, Strong suggested that Barnard's Star would represent the first step along a favourable 'route map' among the stars, with the possibility of future onward connections to stars such as Ross 154 and 24 Ophiuchi. Overall, the possibility of discover-ing new worlds, whether in terms of finding likely conditions for life in the Universe, communicating with extra-terrestrial intelligence, or planning an escape route for doomed humanity, was too enticing an opportunity to ignore in the development of Project Daedalus.

Further explaining the selection of Barnard's Star, Daedalus commit-tee member Tony Lawton reported that the Dutch-American astronomer Peter van de Kamp had in 1969 announced the presence of 'not one, but *two* unseen planetary companions' as part of a wider system around Barnard's Star.[30] Van de Kamp's analysis was based on decades of observations start-ing in 1937 from Sproul Observatory in Pennsylvania, and represents one of the earliest examples of exoplanet detection in astronomy, as outlined in a series of papers from 1963 onwards.[31] The method employed was astrome-try, whereby the observed motion or 'wobble' of a star relative to its back-ground indicates the gravitational effects of an orbiting planet. While Van de Kamp's discovery was later discredited by his own colleagues as a mis-take resulting from instrumental errors, it can in retrospect be seen as an integral part of the scientific process of understanding interstellar space, which, since more accurate techniques have been developed, has led to the discovery of hundreds of exoplanets throughout the galaxy.[32] Van de Kamp's study also represents some of the geographical aspects of profes-sional astronomy in the mid-twentieth century, which was largely reliant on extensive observation regimes at established observatories. As noted by Strong, 'from northern latitudes Alpha Centauri cannot be seen, since it lies close to the Southern Cross' and, therefore, Barnard's Star represents the nearest (if not the brightest) star visible to astronomers working in the northern centres of astronomical science, in Europe and the United States.[33] Indeed, as a relatively dim red dwarf, Barnard's Star is not visible to the naked eye, and its detection therefore relied on telescopic observation and associated technologies. Not having been visible to astronomers of old, it had not been incorporated into any astrological constellations or afforded any significance until its discovery in the early twentieth century by the American astronomer Edward Emerson Barnard (1857–1923). It was noted originally by Barnard for its large relative motion across the skies, appearing to move rapidly against the otherwise fixed stellar backdrop, and was sometimes referred to as 'Barnard's Runaway Star'.[34] It was Van de Kamp's detection of a 'wobble' in this line of movement over many years that led to his assertion of a planetary system.[35]

At the time the BIS was researching Project Daedalus, Barnard's Star was both the nearest star observable to the astronomical community with which the BIS was connected and also the most exciting destination in terms of possible other worlds. Van de Kamp's initial discovery in 1963 was reported in the London *Times*, in an article which outlined the detection of 'a planet one-and-a-half times the size of Jupiter' orbiting the 'second nearest star to the Earth'.[36] With Van de Kamp having claimed a second gas giant in 1969, Lawton was moved to venture the possibility of further planets orbiting Barnard's Star, commenting with increasing levels of detail and certainty:

> There is little doubt that B1 and B2 can be classed as true planets and it is interesting to speculate on their being the major elements of a solar system which has further members [...] Barnard's Star may well have three small inner planets.[37]

Lawton concludes that the star would make an ideal destination for an interstellar mission, not only because of its relative proximity and the possibility of a planetary system, but also because of 'its intriguing resemblance to our own Solar System', which also contains two gas giants and a series of smaller inner planets.[38] As part of his analysis, Lawton juxtaposed a diagram of our Solar System with the extrapolated arrangement of expected planets in the 'Barnard solar system', illustrating the presence of B1 and B2 in an 'ice planet zone', with speculative question marks in an inner 'rock planet zone', mimicking the structure of the familiar home system.

Here, and in the publicity that surrounded Van de Kamp's discovery, the imagined planetary system of Barnard's Star became enrolled in a process of place-making, whereby astronomical data were translated into a belief in the existence of other worlds, by means of imagery, metaphor and analogy. While understanding Van de Kamp's gas giants as 'true planets', the careful use of extrapolation paints a compelling picture of a full solar system. This aggregation of observational data and speculative rendering anticipates broader patterns in the formation of scientific discourses in planetary astronomy, whereby 'planets [...] are imagined as *places* amenable to habitation', either by human or alien life forms.[39]

The theme of life on other worlds was hinted at enticingly towards the end of Lawton's paper, as he contended that 'life may occupy any convenient ecological niche' even though 'the odds are heavily stacked against it in the Solar System of Barnard's Star'.[40] Here, in raising the possibility of extra-terrestrial life, even as an unlikely occurrence, imaginative interest in Barnard's Star was provoked. A later paper by members of the Daedalus committee went somewhat further, interpreting additional data on Barnard's Star and speculating about the possibility of life existing within its planetary system.[41] Here, Anthony Martin and Alan Bond reviewed a series of papers including the latest interventions by Van de Kamp, alongside analyses of other dwarf stars and studies using more detailed spectroscopic methods of deducing stellar and exoplanetary characteristics. They drew various inferences about Barnard's Star, including its life span, the likely formation processes of its planets, as well as possible biological activity on any such planets. This was done by relating the Barnard star system to known processes of stellar and planetary evolution, referring often to the familiar framework of the Earth's Solar System. It was asserted that any planets that formed around Barnard's Star would have done so 'under much lower temperature conditions' than those in our own Solar System, due to the relative sub-luminosity of the star itself.[42] As a result of this process, any smaller inner planets of the Barnard system would contain 'a much larger mass of water, ammonia and hydrogenated carbon compounds' than the Earth.[43] Furthermore, Martin and Bond suggested that, even on the gas giants B1 and B2, temperatures may be 'amenable to the development of higher life forms' in gradated layers within the planetary atmosphere, while prospects for photosynthesis on any inner planets are also strongly hinted at.[44] In writing of a 'stellar ecosphere', and in noting how the expected long age-range of red dwarf systems may have resulted in extended 'evolutionary processes', the Daedalus team present Barnard's Star as a possible harbinger of life in the Universe.[45]

The projected characteristics of Barnard's Star as a destination for Daedalus, including its expected planetary system, were born not out of a purely rational scientific analysis, but drew on a tradition of imagining Barnard's Star as a host for living planetary companions, and a cultural way of seeing Barnard's Star through a sense of mystique or idiosyncrasy that had featured prominently ever since its discovery. A recent update to the story of Barnard's Star has been the discovery of a genuine 'candidate super-Earth' planet orbiting the star by astronomers in 2018, which prompted a series of news articles and artists' impressions describing this likely exoplanet.[46] This kind of scientific imagination of outer space, particularly of other worlds, follows a pattern that can be traced both in respect to recent exoplanet discoveries, and also in enduring understandings of the planets of the Solar System.[47] Indeed, ever since Johannes Kepler in the seventeenth century imagined the Moon as a world in its own right, planetary bodies have been accorded a sense of place and habitability that has remained an integral part of the scientific imagination of outer space.

Daedalus: reconceptualising the Solar System environment

A recent study of cultural responses to a 1994 comet crash on Jupiter described the Earth as 'part of a vast cosmic environment', suggesting that 'environmental history should embrace the whole Universe', starting with the Solar System.[48] This contributes to a growing body of work that conceptualises outer space as part of the wider human environment, both in the context of concerns about the exploitation of outer-space resources, as well as in understandings of the outer-space environment as a realm of conquest or defence.[49] This environmental understanding also implicates ethical frameworks for conceptualising outer-space futures, such that 'human engagement with outer space is also a question of environmental justice', in cases such as the proliferation of orbital debris and the establishment of planetary protection protocols.[50] The speculative nature of many outer-space projects, according to some researchers, does not negate the 'lasting impact' of these ethical frameworks in tangible ways on Earth.[51] What such studies have established is that, instead of seeing the Earth as separate to and exclusive from its cosmic surroundings, our home planet is bound up intimately with the processes and materials of a wider, active Universe.

In many ways, Project Daedalus embraced this notion of the Solar System as part of the Earth's environment, particularly in relation to the assembly of the Daedalus vehicle, the societal conditions under which it could be constructed, and the resources that would be required to set it on its interstellar journey. Hinting at some of these complexities, when asked about whether he actually believed that Daedalus would ever be built at some point in the future, Bob Parkinson responded in an understated fashion that 'it wasn't something you were going to do in a hurry'.[52] Although the project's ethos was to use existing technology wherever possible and only use reasonable extrapolation where necessary, this was still a design that would require enormous resource and engineering capabilities to realise. The Daedalus craft itself would be of substantial size, at 190 metres in length, with a total mass of 450 tonnes, most of which would consist of solid fuel pellets in the form of the radioactive isotopes deuterium and helium-3.[53] With such a great mass, the vehicle would be too large to be built and launched on Earth and would therefore need to be assembled in space, while its fuel resources alone would have to be drawn from across the interplanetary environment. Such considerations fed into the broader aspects of Project Daedalus, enrolling new imaginative geographies of the Solar System that envisaged widespread mining of extra-terrestrial resources, the establishment of advanced outer-space habitats, and an acknowledgement of humankind's active role in colonising the planets.

As part of the final report of Project Daedalus, a series of illustrations by Bill Dillon outlined the construction and launch of the vehicle from the orbit of Jupiter's moon Callisto. In one of these renderings (Fig.6.1), Daedalus takes shape in the upper part of the image, with its globular fuel containers and cone-shaped exhaust section having been put together, while crewed construction craft handle parts of the machinery. The free-floating perspective imagines the viewer of this scene tethered to one of these construction crafts, and a wheel-shaped space station serves as a base of operations, with a crescent-lit Jupiter providing background detail. While reminiscent of the earlier designs of Arthur C Clarke and associated illustrations of R A Smith (see Chapter 4), Dillon's artwork imagines a spacefaring society in which a range of orbiting space habitats, satellites and vehicles form part of a sustained human presence in outer space. Rather than simply

Figure 6.1 Illustration by Bill Dillon showing the construction of the Daedalus vehicle

Source Credit: Bill Dillon/British Interplanetary Society

acting as an illustrative backdrop, these visions of a colonised Solar System of the future became an integral part of Project Daedalus.

The size and resource requirements of Daedalus meant that the design team had to envisage the project as part of a future society that would be capable, technologically, politically and economically, of sustaining a substantial presence in outer space, not only in orbit around the Earth but throughout the Solar System. Summarising this perspective, Bond and Martin explained:

> It seems probable that a Solar System wide culture making use of all its resources would easily be wealthy enough to afford such an undertaking and presumably in order to have reached the stage of extensive interplanetary flight would also have achieved reasonable political stability, and an acceptance of this new environment. [...] We envisage Daedalus-type vehicles being built by a wealthy (compared to the present day) Solar System wide community, probably in the latter part of the 21st century.[54]

Such projections can be seen as part of a particular social and political framework in which futures were imagined in post-war Britain. Explaining this as a type of retro-futurism, Parkinson later described how, as his career in British science and technology developed, a sense of nostalgia for an earlier period of scientific and technological progress often acted as a motivation, while projects stalled and government support for hi-tech industries faltered. In other words, when Parkinson was 'growing up as a teenager the future was a big thing'.[55] In Project Daedalus, therefore, a sense remains that, through the speculative freedom offered by the BIS, the unrealised potential of the Space Age could be brought to life. Indeed, following the milestones of the first artificial satellite and the Moon landings, the prospect of a Solar-System-wide community by the end of the next century seemed logical, and acted as a spur to new ideas. Parkinson anticipated a phase of 'interplanetary transition' in human society in which 'not only the exploration but also the colonization of the Solar System will take place', paving the way for a project such as Daedalus.[56] Scholars in science and technology studies have noted 'the importance of future-oriented discourse in technical practice', particularly in spaceflight technology, outlining a sense in which the future is continuously re-imagined in technical designs.[57] Plans for the future of space exploration within the BIS had long been associated with the colonisation of the Solar System, and this culture of anticipation was certainly an important aspect of Project Daedalus.

The recognition by the Daedalus team of the need for political stability and economic prosperity as part of a 'Solar System wide culture' further codifies the societal parameters in which Daedalus was envisioned. From a macro-economic perspective, Parkinson anticipated an advanced knowledge base across a wide range of technical and scientific sectors, as part of a spacefaring society in which 'men will not only learn the techniques, but also gather the resources for the larger jump to the stars'.[58] As the Daedalus team started to develop the conceptual framework for a global spacefaring culture, it was made implicit that any future society in which Daedalus could exist would be characterised by an extractivist, resource-consuming economy. Studies in environmental geography have demonstrated how conceptualising outer space in terms of natural resource availability has been connected to Earthly discourses of political power and territorial control.[59] This has included the visions of private companies that have proposed mining the Moon and other off-world spaces, while promoting narratives of resource depletion on Earth. Understanding Project Daedalus in such terms, therefore, means acknowledging the political and cultural perspectives from which it was conceptualised, and the lasting impact of such imaginaries in contemporary society.

Also significant in the early 1970s were genuine concerns about global energy provision, with the effects of the 1973 oil shock combining with what Bond and Martin described as the 'hysteria which surrounds nuclear fission', in a likely reference to the anti-nuclear movement.[60] Responding to this discourse of global resource stress, the Daedalus team considered the ways in which the environmental resources of the Solar System might enable solutions to the world's energy problems. As part of these considerations, it was assumed that, 'in the next century helium-3 from Jupiter may already be returned to Earth on a routine basis for consumption in ground-based fusion reactors'.[61] Nuclear fusion was, and still is, seen by some as a panacea for the future energy requirements of a technologically advanced population, while alleviating some of the fears about nuclear waste that had come to dominate public discourse on atomic power by the 1970s. Nuclear fusion using helium-3 was valued by the Daedalus team not only for these reasons, but also as the best option for powering the Daedalus vehicle itself. Following the principles of Project Orion, the propulsion mechanism for Daedalus would involve igniting a pellet stream of helium-3–deuterium isotopes with a high-powered electron beam, with the resultant small fusion explosions propelling the spacecraft in a series of consecutive pulses.[62] The main problem with this design was that helium-3 is virtually non-existent on Earth, and would have to be acquired from elsewhere in the Solar System. In an oddly circuitous forecast, the Daedalus spacecraft would be reliant on extraction of helium-3 from across the Solar System, while the imagined society in which Daedalus could exist, would likely have developed extensive mining capabilities throughout the Solar System already.

Central to this set of circumstances was the concept of extracting helium-3 and deuterium isotopes in large quantities from the atmosphere of Jupiter.[63] This problem was approached by Bob Parkinson through three possible solutions, with helium-3 being the most pressing concern, due to its relative scarcity. The first option was to manufacture the isotope through the breeding and radioactive decay of tritium, a process so environmentally damaging that banishing any such production plant to the Moon would have to be considered. The second option was to collect helium-3 from the solar wind, particles that are ejected from the Sun's corona to interplanetary space, but this was also deemed impractical. The preferred option was to collect both isotopes from the atmosphere of Jupiter, where they were believed to be abundant enough to supply the required 30,000 tonnes of helium-3 and 20,000 tonnes of deuterium. Here, adequate supplies could be harvested by an array of collecting balloons over a period of twenty years, and transported to a processing station orbiting one of Jupiter's moons, ready to be loaded onto the Daedalus vehicle. These 'aerostat factories' would float within the Jovian atmosphere and process a combined total of 28 tonnes of gas per second.[64] In imagining and seriously thinking through this process, Parkinson was able to put in place one of the pieces of the puzzle for Project Daedalus.

Running through Parkinson's paper on propellant acquisition were detailed descriptions and analyses of the 'geographies' of the planet Jupiter itself. With a series of static balloons central to the isotope extraction plan, a detailed understanding of the Jovian atmosphere was required, along with an appreciation of factors including the gravitational effects of such a large planet, its rate of rotation and the constitution of its weather patterns. Here, NASA's Pioneer programme was directly influential, with photographs of Jupiter in unprecedented detail being taken from the fly-pasts in 1973 and 1974 and released to the public. These encounters represented the very first exploratory missions of the Jovian system, and greatly enhanced human knowledge about the planet. This included details of the Jovian cloud systems, as well as data on the planet's magnetosphere and radiation emissions, demonstrating that 'spacecraft could explore Jupiter and survive the hazards of the Jovian environment'.[65] This new knowledge was available to Parkinson at the time he was planning his contributions to Project Daedalus, and one of Pioneer 11's photographs of Jupiter illustrates his report, showing various weather patterns including the famous Red Spot. Parkinson interprets this image, suggesting that 'the visible zones and belts circling the planet [...] are the product of Coriolis winds driven by vertical convection currents', resulting in 'wind speeds of up to 90 m/sec', a hazard for any exploratory mission into Jupiter's volatile atmosphere.[66] Here, knowledge of terrestrial weather systems and broader cosmography were extrapolated in order to understand Jupiter's complex nature, forming part of the essential knowledge base of Project Daedalus.

The final report of Project Daedalus consisted of twenty-one research papers, at least eight of which were devoted to understanding the outer-space environment in which Daedalus was imagined, including Barnard's Star, the terrestrial Solar System and the spaces in-between. Even the more technical papers, including studies on propellant types, navigational techniques and spacecraft structure, had environmental, cultural and political aspects. Interpreting these dimensions demonstrates how Daedalus was by no means an isolated project, purely confined to engineering or technical specifications. In fact, it required a full deployment of imagined future scenarios for space exploration, and would be reliant on understanding the Solar System in environmental terms. This incorporated the latest knowledge about Jupiter's atmospheric make-up and planetary environment, drawing from the achievements of the Pioneer missions that had been a strong influence on the project as a whole. It was perhaps the comprehensive nature of Project Daedalus that has led it to acquire an iconic status in the history of spaceflight design, being appreciated not just for its technical speculations, but also for imagining, in a somewhat hopeful sense, what could be achieved if humankind could work together in outer space, perhaps in opposition to the geo-strategic or neo-colonialist framework that arguably characterised earlier collaborative efforts, as recounted in Chapter 5. At the same time, however, the synthesis of environmental, economic and political understandings of the Solar System that Daedalus invoked were ultimately reflective of, and continued to inspire, certain ways of imagining outer space as a realm for human exploitation.

Interpreting Daedalus: cultural and scientific impressions

After the final report of Project Daedalus was published in 1978, the Daedalus committee was able to publicise its findings in various ways, reaching out to the wider cultural institutions of science, technology and outer space in Britain. Project Daedalus also stimulated significant work within the BIS on 'Interstellar Studies', representing a new focus within the Society on understanding spaces beyond the Solar System. Some thirty years later, the BIS returned to the concept of an interstellar mission with Project Icarus, a study that has aimed to update the Daedalus concept, taking advantage of more recent scientific and technical knowledge. By interpreting these projects in the period during and after Project Daedalus was undertaken, a sense of its scientific and cultural impact can be ascertained.

One of the ways in which Project Daedalus connected with public audiences and interacted with diverse cultures of creative practice was through the medium of science television. Since the Second World War, the BBC in particular had been experimenting with new types of television broadcast with a popular science focus, including programmes such as *The Sky at Night, Horizon* and *Tomorrow's World.* As Timothy Boon has shown, television 'revolutionised the visual representation of science and vastly expanded its audience', and with the establishment of BBC2 in 1964, science programming grew once again to incorporate new forms of expertise and cultivate audiences.[67] Furthermore, researchers have shown how the science of outer space was enrolled as an important aspect of post-war broadcasting in Britain, both in terms of the technologies of production and transmission that it enabled, as well as in its capacity to appeal as a fundamentally visual science.[68] With this expansion in science television came a raft of new broadcasting professionals specialising in science programming. One such person was Nigel Calder, who came to prominence as a television writer having previously worked as a writer and editor for *New Scientist* magazine. His 1978 three-part documentary series for BBC2, *Spaceships of the Mind*, covered topics including asteroid mining, the colonisation of space, and artificial intelligence, blending philosophical and scientific approaches to understanding the cosmos. The programme's production notes stated that the series would 'examine how recent discoveries in fundamental research may alter human ideas, values and aims'.[69] Moreover, the programme's title acted as a metaphor for 'big ideas' in science, aligning the concept of space exploration with notions of radical change in society. In an accompanying book, Calder stated:

> Because we live at a special moment of scientific history our generation is able to launch spaceships of the mind that may serve as pathfinders for millions of years.[70]

Indeed, according to Calder 'the most radical and hopeful ideas' that emerged in the research for the series 'clustered around the implications of spaceflight'.[71] Some of these concepts were outlined over the course of the three programmes, including Princeton physicist Gerard K O'Neill's designs for advanced space habitats and Freeman Dyson's notion of a 'swarm of orbiting platforms' that 'intercept nearly all the radiant energy of the parent star', that became known as the Dyson Sphere.[72] Alongside such iconic concepts of space colonisation was added Project Daedalus.

Featuring an interview with Alan Bond, the third part of *Spaceships of the Mind*, entitled 'Seeding the Universe', included a section on Project Daedalus. Here, Bond maintained the biological metaphor of the programme's title, framing Daedalus as the first phase in the eventual colonisation of the Universe, with humankind 'spreading something like mould on a wall', echoing Wellsian notions of life in outer space (see Chapter 2).[73] One innovative facet of the new science television programming was the incorporation of visual effects, including props, backdrops and lighting techniques. To support the narrative of the Daedalus mission, the producers of *Spaceships of the Mind* were able to draw on an amateur tradition of craft and visual representation that had long been a feature of the British Interplanetary Society. In the BIS, practices such as model-making and astronomical painting helped establish 'the presence and legibility of space exploration and astronautics within many different contexts including science, politics and popular culture'.[74] As such, models including a replica of the Daedalus ship were constructed for *Spaceships of the Mind* by designer Mat Irvine, who himself was a member of the BIS.[75] Following discussions with Alan Bond, the only addition Irvine made to the design was the name 'Daedalus' on the side of the model, which was otherwise true to the designs of the Daedalus committee.

As a further part of the visual effects for *Spaceships of the Mind*, 'a twenty-foot long black backing, punctured with back-lit stars' was used, with the Daedalus model suspended with invisible wiring.[76] In addition, paintings by space artist David Hardy of 'Saturn in close-up [and] Barnard's Star system' were deployed, as the imagined journey through the interstellar environment was brought to life.[77] As part of this visual effects suite, the technique of front axial projection was used, whereby a still or moving background image is projected onto a reflective screen via an angled mirror, a technique that pre-figured contemporary 'green screen' technology for rendering background scenarios on film. Script notes for one sequence indicate Daedalus 'turning and showing engine firing' to a score of Alan Hovhannes' *Odysseus Symphony*.[78] The association of space exploration with classical music channels a sense of grandeur and the sublime, as can be witnessed in examples such as the introductory music to *The Sky at Night*, Jean Sibelius' *Pelléas et Mélisande*, from 1957, as well as Johan Strauss II's *The Blue Danube* alongside other classical pieces in Kubrick and Clarke's feature film *2001: A Space Odyssey*, in 1968. In any case, the enrolment of music, visual art and craft in cultures of television production, alongside the technical and scientific designs of Project Daedalus, demonstrates how the imagined places of interstellar exploration were created through inter-disciplinary creative endeavour.

It was not just in the popular media that Project Daedalus stimulated new work. Within the BIS itself, a new special series of the *BIS Journal* on 'Interstellar Studies' was published from 1974, becoming known as the 'red cover' editions, running until 1991. During this period, three issues a year were devoted to this new aggregation of themes in space science, and, writing in an editorial to the first issue, Bond and Martin outlined the expected topics that would be explored:

> [T]he possibilities for the existence of intelligence in the Universe, development of methods of contact via radio or other means, work on interstellar propulsion methods and spacecraft systems, development of the capabilities necessary to conduct scientific research at interstellar distances, the formation of planetary and stellar systems, their evolution and means of detecting them and biology and biochemistry concerned with the evolution of life.[79]

This manifesto effectively broadened the scale of space science from the interplanetary to the interstellar, anticipating a step-change in the scientific contemplation of time and space. A sample of the 'Interstellar Studies' papers reveals a combination of the themes set out by Bond and Martin, with some coalescence around the topics of extra-terrestrial life, the extra-Solar spaces of the Galaxy, and various concepts of interstellar flight, with contributions from BIS members in the UK, but also from researchers in the United States and continental Europe. As such, P Molton of the University of Maryland's Chemistry Department pondered the possibilities of 'Non-Aqueous Biosystems', and the potential of ammonia to act as the central element of life on other worlds, questioning the very chemical basis of life as we know it.[80] Furthermore, as part of a special issue on 'World Ships', Gregory L Matloff of City University New York outlined possibilities for the far-future migration of humankind to interstellar space, with the Sun having reached its 'White Dwarf' phase of stellar evolution.[81] The special issue as a whole expanded on Leslie Shepherd's prior concept of the 'generation ship' to envisage a wholesale exodus of humanity away from a dying Sun. Some papers chose to imagine the conditions of other worlds and star systems, with F D Seward of the University of California asking readers to 'assume that the Earth is transported to a point inside the Crab Nebula'.[82] From this point, Seward 'compare[s] the environment there with the environment here', describing the formidable magnetic and radiative conditions that would arise in the presence of the Crab Pulsar that lies at the centre of this Nebula.[83] In the Interstellar Studies papers, researchers and readers were able to cast their scientific imaginations out into the cosmic environment in ways that had not been done before. The BIS acted as a conduit for information flowing between the United States and the UK around this topic, enhancing its journal's reputation as one of the world's pre-eminent astronautics publications. The influence of Project Daedalus thereby came to be felt by a generation of space enthusiasts.

In 1986, Bond and Martin re-visited Project Daedalus in a retrospective article. They claimed that 'a reasonably convincing case had been made for the feasibility of interstellar probes' and that 'much grander interstellar missions were possible'.[84] An important implication of this was that Fermi's Paradox could be considered unresolved, and the question of whether or not life exists elsewhere in the Universe could remain open for further debate. Daedalus continues to influence activity within the BIS, and in the late 2000s, the Society established Project Icarus ('Son of Daedalus').[85] This was to be a successor interstellar design study, using Daedalus as a starting point and one of the project's early ventures was to commission a new Daedalus model to put on display in the BIS headquarters in South London (Fig. 6.2). Project Icarus has since morphed into the non-profit foundation Icarus Interstellar, an international consortium of researchers pursuing a range of projects, with the goal of realising interstellar travel by the year 2100. Alongside other initiatives such as NASA and DARPA's '100 Year Starship' grant project and the Tau Zero Foundation, Icarus Interstellar seeks to mobilise citizen scientists, professional researchers and space entrepreneurs to establish a new generation of individuals working towards the goal of interstellar flight. Its founder Andreas Tziolas has explained some of the motivations behind Project Icarus, which include ensuring the 'survival of humankind', alongside the desire to 'push technological boundaries', acknowledging also the influence of Project Daedalus.[86] In bringing together networks within and outside of the BIS, Icarus represents a looser

Figure 6.2 Daedalus model (second engine stage) by Terry Regan, commissioned
  by the BIS in 2011

Source Credit: British Interplanetary Society

and less-focussed design project than Daedalus, but nevertheless demonstrates the enduring fascination with interstellar flight, both in Britain and
internationally, that Daedalus had magnified thirty years previously.

Conclusion: geographies of interstellar space

In turning to the spaces beyond the Solar System and the possibilities that
arose from Project Daedalus, this chapter has explored the ways in which
a different set of geographical concepts can be used to understand the cultures and politics of outer space in Britain. Ideas of place were enrolled
by those who sought to imagine spaces such as the planetary companions
of Barnard's Star as destinations and as possible harbingers of life in the
Universe. New concepts of exploration were also anticipated by those considering interstellar voyages, both in terms of scientific missions to the stars
as well as the prospect of generation-ships that would in the far-future take
humankind to new homes in the cosmos. Practical and ethical considerations of the Solar System as part of Earth's cosmic environment were also
developed by members of the Daedalus committee who anticipated the
resourcing and construction of the immense Daedalus vessel. Here, close
understandings of the geographies of Jupiter were enrolled, drawing from
the stunning achievements of the Pioneer space probes. This chapter has

also contemplated the ways in which Project Daedalus was connected to Earthly cultural and political discourses, whether through the military-industrial programmes that provided the intellectual capital to pursue such projects, or in terms of the popular cultural appeal of the Daedalus concept. As such, Project Daedalus has provoked a number of engagements with the geographies of outer space in Britain during the mid-twentieth century, placing cultures of interstellar flight here on Earth while also establishing imaginative geographies of the Solar System and beyond. While Daedalus undoubtedly advanced such conceptual engagements with space exploration in Britain, the late twentieth century witnessed an expansion of activity in outer space, including a number of space missions, both human and non-human, that in many ways matched the dreams of earlier space visionaries.

Notes

1 Alan Bond and Anthony R Martin, 'Project Daedalus', *JBIS Supplement* (1978), 5–7 (p.6–7) [emphasis in original].
2 Lisa Messeri, *Placing Outer Space – an Earthly Ethnography of Other Worlds* (London: Duke University Press, 2016); Dagomar Degroot, '"A Catastrophe Happening in Front of Our Very Eyes": The Environmental History of a Comet Crash on Jupiter', *Environmental History*, 22 (2017), 23–49; Julie Klinger, *Rare Earth Frontiers – from Terrestrial Subsoils to Lunar Landscapes* (Ithaca: Cornell University Press, 2017).
3 Eric M Jones, *"Where is everybody?" An account of Fermi's question* (Los Alamos: Los Alamos National Laboratory, 1985) https://fas.org/sgp/othergov/doe/lanl/la-10311-ms.pdf [31st July 2020].
4 NASA / Jet Propulsion Laboratory, 'Voyager Mission Status', *NASA / Jet Propulsion Laboratory* (2020) https://voyager.jpl.nasa.gov/mission/status/ [10th August 2020].
5 Eric Burgess, *By Jupiter – Odysseys to a Giant* (New York: Colombia University Press, 1982), p.145.
6 Burgess, *By Jupiter* (p.146).
7 Carl Sagan, Linda Salzman-Sagan, and Frank Drake, 'A Message from Earth', *Science*, 175 (1972), 881–884 (p.884).
8 William Macauley, 'Inscribing Scientific Knowledge: Interstellar Communication, NASA's Pioneer Plaque, and Contact with Cultures of the Imagination, 1971–1972', in *Imagining Outer Space – European Astroculture in the Twentieth Century*, ed. by Alexander Geppert (Basingstoke: Palgrave Macmillan, 2007), pp. 285–303.
9 NASA / Jet Propulsion Laboratory, 'Voyager – The Golden Record', *NASA / Jet Propulsion Laboratory* (2020) https://voyager.jpl.nasa.gov/golden-record/ [10th August 2020].
10 Leslie Shepherd, 'Interstellar Flight', *Journal of the British Interplanetary Society,* 56 (1952), 80–91 (p.80).
11 James E Strong, *Flight to the Stars* (London: Temple Press Books, 1965), p.ix.
12 John Krige, 'Atoms for Peace, Scientific Internationalism and Scientific Intelligence', *Osiris,* 21 (2006), 161–181; Fraser MacDonald, 'Space and the Atom: On the Popular Geopolitics of Cold War Rocketry', *Geopolitics,* 13 (2008), 611–634.

13   Nuclear energy has also since been applied to power electrical circuits on deep space probes such as Pioneer 10 and 11, in the form of radioisotope thermoelectric generators.

14   Cornelius Joseph Everett and Stanislaw Marcin Ulam, *On a Method of Propulsion of Projectiles by means of External Nuclear Explosions* (Los Alamos Scientific Laboratory: Defense Technical Information Center, 1955).

15   *To Mars by A-Bomb: The Secret History of Project Orion,* Produced by Christopher Sykes (BBC, 2009).

16   George Dyson, *Project Orion – The Atomic Spaceship, 1957–1965* (New York: Henry Holt, 2002).

17   The basic principles of Project Orion were declassified in October 1964, and Dyson wrote about its cancellation in Freeman J Dyson, 'Death of a Project', *Science*, 149, 3680 (1965), 141–144.

18   Danny B Stillman and Thomas C Reed, *The Nuclear Express : A Political History of the Bomb and Its Proliferation* (Minneapolis: Zenith Press, 2010).

19   'Leslie Shepherd', *Telegraph*, 16th March 2012, Science Obituaries.

20   Shepherd, 'Interstellar Flight' (p.80).

21   Shepherd, 'Interstellar Flight' (p.83).

22   Olaf Stapledon, 'Interplanetary Man?', *Journal of the British Interplanetary Society,* 7 (1948), 213–233 (p.230).

23   Shepherd, 'Interstellar Flight' (p.81).

24   Shepherd, 'Interstellar Flight' (p.84).

25   *Limiting Outer Space*, ed. by Alexander Geppert (London: Palgrave Macmillan).

26   David Edgerton, *Warfare State: Britain, 1920–1970* (Cambridge: Cambridge University Press, 2006).

27   Thomas Lean, *National Life Stories – An Oral History of British Science. Dr Bob Parkinson, Interviewed by Dr Thomas Lean, 2010–2011, C1379/05* (The British Library) https://sounds.bl.uk/related-content/TRANSCRIPTS/021T-C1379X0005XX-0000A0.pdf [31st July 2020].

28   Alan Bond and others, 'Project Daedalus – the Final Report on the Bis Starship Study', *JBIS Supplement* (1978), 5–8.

29   James E Strong, 'Which Highway to the Stars?', *Spaceflight,* 12 (1970), 174–177.

30   Anthony T Lawton, 'The Nearest Other Solar System?', *Spaceflight,* 12 (1970), 170–173 (p.170) [emphasis in original].

31   Peter Van de Kamp, 'Astrometric study of Barnard's star from plates taken with the 24-inch Sproul refractor', *The Astronomical Journal*, 68 (1963), 515–521.

32   Athena Constemis and Thérèse Encrenaz, *Life Beyond Earth – the Search for Habitable Worlds in the Universe* (Cambridge: Cambridge University Press, 2013).

33   Strong, *Flight to the Stars* (p.67).

34   William Sheehan, *The Immortal Fire Within – The Life and Work of Edward Emerson Barnard* (Cambridge: Cambridge University Press, 1995), p.402.

35   Barnard's Star continues to hold the record for the largest relative motion of any visible star, and will in several thousand years become the closest star to the Earth other than the Sun.

36   'US Discovery of New Planet', *The Times*, 19th April 1963 (55679): 12.

37   Lawton, 'The Nearest Other Solar System?' (p.170, p.172).

38   Lawton, 'The Nearest Other Solar System?' (p.173).

39   Messeri, *Placing Outer Space* (p.9).

40   Lawton, 'The Nearest Other Solar System?' (p.173).

41   Anthony R Martin and Alan Bond, 'Project Daedalus: An Analysis of the Photometric Data on Barnard's Star, and Its Implications', *JBIS Supplement* (1978), 24–32.

42   Martin and Bond, 'Project Daedalus' (p.30).

43 Martin and Bond, 'Project Daedalus' (p.30).

44 Martin and Bond, 'Project Daedalus' (p.30).

45 Martin and Bond, 'Project Daedalus' (p.30, p.31).

46 Ignasi Ribas and others, 'A candidate super-Earth planet orbiting near the snow line of Barnard's star', *Nature* 563 (2018), 365–368.

47 Messeri, *Placing Outer Space.*

48 Degroot, 'A Catastrophe Happening', (p.23).

49 Valerie Olson, 'Political Ecology in the Extreme: Asteroid Activism and the Making of an Environmental Solar System', *Anthropological Quarterly,* 85 (2012), 1027–1044; Everett C Dolman, *Astropolitik: Classical geopolitics in the space age* (London: Frank Cass, 2002).

50 Julie Klinger, 'Environmental Geopolitics and Outer Space', *Geopolitics* (2019), 1 [online].

51 Matthew Kearnes and Thom van Dooren, 'Rethinking the Final Frontier: Cosmo-Logics and an Ethic of Interstellar Flourishing', *Geohumanities*, 3 (2017), 178–197 (p.187).

52 Thomas Lean, *National Life Stories.*

53 James E Strong and Alan Bond, 'Project Daedalus: The Vehicle Configuration', *JBIS Supplement* (1978) 90–95.

54 Alan Bond and Anthony R Martin, 'Project Daedalus', *JBIS Supplement* (1978): 5–7 (p.7).

55 Thomas Lean, *National Life Stories.*

56 Bob Parkinson, 'The Starship as an Exercise in Economics', *Journal of the British Interplanetary Society,* 27 (1974), 692–696 (p.696).

57 Lisa Messeri and Janet Vertesi, 'The Greatest Missions Never Flown – Anticipatory Discourse and the "Projectory" in Technological Communities', *Technology and Culture,* 56 (2015), 54–85.

58 Parkinson, 'The Starship as an Exercise in Economics' (p.696).

59 Klinger, *Rare Earth Frontiers.*

60 Bond and Martin, 'Project Daedalus' (p.6).

61 Bond and Martin, 'Project Daedalus' (p.7).

62 Anthony R Martin and Alan Bond, 'Project Daedalus: The Propulsion System', *JBIS Supplement* (1978), 45–61.

63 Bob Parkinson, 'Project Daedalus: Propellant Acquisition Techniques', *JBIS Supplement* (1978), 83–89.

64 Parkinson, 'Project Daedalus: Propellant Acquisition' (p.85).

65 Burgess, *By Jupiter* (p.43).

66 Parkinson, 'Project Daedalus: Propellant Acquisition' (p.87).

67 Timothy Boon, *Films of Fact: A History of Science in Documentary Films and Television* (London: Wallflower, 2008), p.184.

68 James Farry and David A Kirby, 'The Universe Will Be Televised: Space, Science, Satellites and British Television Production, 1946–1969', *History and Technology,* 28 (2012), 311–333.

69 Reading, Caversham Park, BBC Written Archives Centre, 'Spaceships of the Mind – General', T63/148/1.

70 Nigel Calder, *Spaceships of the Mind* (London: British Broadcasting Corporation, 1978), p.141.

71 Calder, *Spaceships of the Mind* (p.8).

72 Calder, *Spaceships of the Mind* (p.22).

73 Reading, Caversham Park, BBC Written Archives Centre, 'Spaceships of the Mind' Final Script, Programme 3, 'Seeding the Universe', 5th July 1978, BBC2 (p.19) [on microfilm].

74 William Macauley, 'Crafting the Future: Envisioning Space Exploration in Post-War Britain', *History and Technology,* 28 (2012), 281–309 (p.282).

75 Reading, Caversham Park, BBC Written Archives Centre, 'The Making of Spaceships of the Mind', T63/148/1.
76 'The Making of Spaceships of the Mind', T63/148/1.
77 'The Making of Spaceships of the Mind', T63/148/1.
78 'Spaceships of the Mind' Final Script, Programme 3, 'Seeding the Universe', 5th July 1978, BBC2 (p.18).
79 Anthony R Martin and Alan Bond 'Editorial', *Journal of the British Interplanetary Society*, 27 (1974), 241.
80 Peter Molton, 'Non-Aqueous Biosystems: The Case for Liquid Ammonia as a Solvent', *Journal of the British Interplanetary Society*, 27 (1974), 243–262.
81 G L Matloff, 'World Ships and White Dwarfs', *Journal of the British Interplanetary Society* 39 (1986), 114–115.
82 F D Seward, 'A Trip to the Crab Nebula', *Journal of the British Interplanetary Society*, 31 (1978), 83–92.
83 Seward, 'A Trip to the Crab Nebula' (p.83).
84 Alan Bond and Anthony R Martin, 'Project Daedalus Reviewed', *Journal of the British Interplanetary Society,* 39 (1986), 385–390 (p.385).
85 Kelvin F Long and others, 'Project Icarus – Son of Daedalus – Flying Closer to Another Star' *Journal of the British Interplanetary Society* 62 (2009), 403–414.
86 Ross Andersen, 'Project Icarus: Laying the Plans for Interstellar Travel', *The Atlantic*, 23rd December 2012.

7 Space exploration, science and nationalism

Concepts of progress in outer space had, towards the end of the twentieth century, become more complex and multi-faceted than in the preceding era of giant leaps and iconic first achievements. Increased international co-operation in space, the heightened role of private finance in supporting space endeavours, and the emerging potential of robotic probes to explore the planets of the Solar System had all challenged and changed the complexion of space exploration. Collaborative ventures were encouraged by the Apollo-Soyuz test project in 1975, while the same year saw the formation of the European Space Agency, which began to emerge as a major player in space exploration. Towards the end of the 1980s, the changing geopolitical situation carried further implications for outer-space activities.[1] Most significant was the end of the Cold War in 1991, which preceded a relaxation of national rivalries in outer space.[2] Thereafter, plans for an American space station named 'Freedom' to rival the Soviets' *Mir* morphed into designs for an international space station, with Russian, American, European and Japanese involvement. Similarly, the emergence of private companies as increasingly powerful actors in space following the success of a range of commercial satellites in the 1960s can be seen as a further riposte to the supremacy of national agency in space. Following a discussion on theories of nationalism and the ways in which they have been applied to activities in outer space, this chapter focusses on three British space missions in the 1990s and 2000s: Helen Sharman's spaceflight experience on board *Mir* in 1991, the 'Beagle 2' Mars lander project in 2003, and Tim Peake's mission to the International Space Station in 2015–16. While each of these missions involved British collaboration with other national and international space agencies, this chapter suggests that concepts of nationalism and national identity retained a key role in space exploration in the years either side of the new millennium.

Nationalism and cultures of outer space

Concepts of nationalism have formed an active component of critical political geographies for at least the past forty years, and many scholars now agree that nationalism remains a powerful ideological force in today's world.

This growing consensus has emerged contrary to schools of thought that have downplayed the role of nations in a postmodern, globalised world. Indeed, researchers have been encouraged to consider nationalism on an equal footing to its 'ideological competitors' socialism and liberalism, in recognising the widespread effects it has on individuals and societies at large.[3] Accordingly, scholarly work in human geography and allied disciplines has started to configure new approaches to understanding nationalism's continued allure, in some cases looking towards outer space for examples of the ways in which nationalism has been expressed and understood.

While some scholars have considered nations as 'imagined communities' drawing from deep cultural roots and limited by geographical boundaries, others have pointed towards the material, tangible aspects of nationhood to which individuals often see themselves as tied.[4] Concepts of everyday or banal nationalism run counter to the expectation that nationalism occupies only fringe or extreme social movements in areas peripheral to Western developed nations. Nationalistic sentiments have also been accounted for outside the boundaries of a given nation-state, with diasporic and remote associations recognised alongside home-grown iterations. Such debates have progressed by taking account of the daily, often unnoticed expressions of nationhood that have been seen as 'endemic' in contemporary societies, with the unwaved flag becoming the ubiquitous symbol of banal nationalism.[5] Drawing from this work, geographers have increasingly focussed on popular culture and the ordinary, shared experiences of members of nations, bringing to bear a multitude of new sources and methodologies in understanding nationalism.[6]

That the exploration of outer space has been connected to concepts of nationalism should come as no surprise to those familiar with the photographs of the Apollo moon landings featuring the American flag, or the equally propagandistic promotion of the Soviet space programme as a symbol of national achievement at the height of the Cold War. The United States space programme has been associated with the foundational myths of the American nation, while sources including science-fictional landscape paintings have been understood as central to the promotion of technologically deterministic ambitions for national space leadership in the post-war period. In this way, America is imagined through space exploration as a 'transcendental state', embodying the value-laden concepts of American exceptionalism, frontier expansionism and manifest destiny.[7] A range of detailed studies have connected American nationalism to its space programme, including understandings of outer space as one of the 'frontiers for the American Century', and investigations into the connections between popular culture, spaceflight and American national identity.[8] While there is an established literature on the cultural and political significance of American space exploration, new space programmes in developing countries offer further opportunities for researching nationalist agendas. In the past few years, for example, states such as China, India and Israel have sought to showcase their national technological capacities through achievements in space exploration, supporting a broader set of geopolitical ambitions and drawing from a corresponding set of cultural agendas and political imperatives.

Offering a renewed conceptual focus on nationalism, geographical researchers drawing from non-representational theory have explored the ways in which encounters between places, bodies and objects have helped to generate feelings of nationhood, such that 'national belonging' becomes 'a felt force in the moment a body's experience encounters foreshadowing, memories, imagined practices'.[9] As such, nationalism can be thought of not just through visual symbols or artefacts of material culture, but as a set of emotions that are personal and relational. Through such approaches, a diverse set of engagements with nationalism has emerged in cultural and political geography, incorporating both material objects as well as affective atmospheres.[10] Approaching nationalism in this way has allowed researchers to investigate the interplay between space artefacts and the general public, particularly in national museum spaces such as those of the Smithsonian Institution in the United States, whether in terms of the lasting iconography of spaceflight mission insignia, or through embodied experiences at Space Shuttle display exhibitions.[11] In this way, the fleet of retired Space Shuttles takes on a new life as a 'body within a constructed assemblage of remembering human spaceflight' as part of a wider 'national affective atmosphere' in the United States.[12] Here is a sense in which the afterlife of space hardware still resonates in the popular consciousness of nationalism, re-enforcing earlier cultural and political discourses associated with the US space programme at the peak of its powers.

Further exploring the embodied aspects of nationalism, feminist scholars have acknowledged that 'the nation cannot be discussed in gender-neutral terms', moving to the scale of the body to examine notions including reproductive citizenship, pageantry in national beauty contests, and other gendered performances.[13] In doing this, a variety of uneven power relations are brought to light in the study of nationalism, exposed through the ways in which gender is enrolled in the public sphere, sometimes through acts of violence. While studies in political geography have mostly focussed on the female body as a contested space of nationalism, the male body has also been associated with nationalist discourse in a variety of different ways. With the advent of human spaceflight in 1961 came a renewed public focus on space exploration amid heightened geopolitical tensions, as the first men in space came to embody the wider geopolitical ambitions of competing Cold War nations. Researchers have shown how the Soviet space programme in the early 1960s was shaped around Yuri Gagarin as a 'massive propaganda juggernaut', while the Communist Party's stipulation that the first cosmonaut would necessarily be from 'a completely Russian and working class background' became an important part of Gagarin's public image.[14] Subsequent to his pioneering journey into space, Gagarin came to embody the ideals of the archetypal Soviet citizen, becoming one of the select 'mythologiz[ed] cosmonauts' who were presented as icons of the national contest with the West for scientific and technological superiority.[15] In the American context, similar tropes have been identified in the discourse that surrounded the 'Mercury Seven' astronauts in the pre-Apollo era. These men were put forward by NASA into the public sphere to be 'eulogized as beacons of the bodily regime required to organize America's exceptional destiny' in outer space.[16] Along with a core group of post-war test pilots in the US Air Force, they were seen as embodying characteristics such as tolerance of risk, and stoicism in the face of physical hardship, later being categorised as enigmatically having 'The Right Stuff' of American legend.[17] Along with this sense of embodied nationalism in male astronauts, the astronauts' wives were reciprocally constructed in their domestic roles as passive, supportive, heteronormative bodies, through their profiling in various media outlets. Through these examples, it is apparent that embodied nationalism takes on a particular resonance in outer-space cultures, and that gender has played an important role in these processes.

Looking at British outer-space endeavours since the mid-twentieth century promises a different and possibly more nuanced set of readings into the ways in which national identity and nationalism may be connected to the extraordinary, or perhaps routine, activities associated with space exploration. For example, the British acquisition of the American 'Corporal' missile in the 1950s has been understood in the context of the UK's desire to hold a stake in the central geopolitical affairs of the post-war period, adding a vertical dimension to projections of national prestige.[18] As such, the Corporal propagated notions of national defence and military power, not only in the eyes and minds of military and political personnel that witnessed the rocket's test-launch from Uist in the Outer Hebrides, but also in the ludic experiences of a generation of youngsters who collected Corporal toys and trading-cards, as well as through the spectacle of the rocket's display in prominent public locations such as Glasgow's George Square.[19] Further studies into the nascent British space programme have associated national prestige with ballistic missile technologies, nuclear power and outer-space research, and Chapter 5 of this book has explored the ways in which nationalistic sentiment played a role in the geopolitical cultures that permeated programmes such as the Ariel satellite system. Looking at the more recent past, and the era of human spaceflight, it becomes possible to examine the ambitious plans for space exploration that have since been carried out. The remainder of this chapter synthesises questions of nationalism with British programmes of space exploration since the early 1990s, specifically the space missions of Helen Sharman, Beagle 2 and Tim Peake. It argues that the interrelated concepts of embodied nationalism, affective nationalism and banal nationalism are woven through each of these case studies, and thereby positions nationalism as one of the key components of the geographies of outer space in Britain.

Helen Sharman: banal nationalism
and narratives of exploration

Helen Sharman's personal story became an integral part of the narrative of the first Briton to go into space. Perhaps most remarkable was the fact that prior to her training at the Soviet space centre 'Star City', Sharman had no experience in space science, aeronautics, or the armed forces, characteristics which had been central to the profiles of most astronauts and cosmonauts since the advent of the Space Age. Indeed, from 1984 to 1986, four British military and defence department personnel had gone through astronaut training in the United States as Space Shuttle payload specialists as part of the Skynet military satellite programme, a project that was abandoned after the Challenger disaster in 1986.[20] Sharman broke the mould, not least because she was a woman, and as she later recounted:

> I did not know I was growing up to do the job I did. I did not even aspire to do that job. When the opportunity of the job came up, though, I was, quite unexpectedly, ready to do it.[21]

Sharman's unconventional route into space began in June 1989, the moment she heard a radio advertisement for 'Project Juno' while driving home from work in Slough, on the outskirts of London. One widely recounted part of this story was the tagline of the radio ad, which idiosyncratically stated: 'Astronaut wanted – no experience necessary'.[22] Having scribbled down the phone number while waiting at traffic lights, Sharman made the call after a few days, barely giving it much serious thought at all. With her background in chemistry, she had been working at the confectionary giant Mars for a couple of years, enjoying the satisfaction that came with applying her specialist knowledge of food science to a fulfilling role with good career prospects. Indeed, she applied to Project Juno more as an exciting personal development opportunity rather than to be involved in a historic moment of national achievement.[23]

Project Juno was a collaborative venture between the Soviet space agency *Glavkosmos* and the British private company Antequera Ltd, involving aspects of popular science, public-private partnerships and public relations in its goal of sending a British visitor to the *Mir* space station. Around this time, *Mir* had started welcoming high-profile paying visitors from non-Soviet nations, becoming characterised as a 'kind of space hotel'.[24] Recognising this opportunity, Project Juno was co-founded by the German-British scientist Heinz Wolff, who had become known as a television presenter and populariser of science in Britain during the 1970s and 80s.[25] The project's founders also understood that the recruitment competition would feed into the publicity machine, and set the eligibility criteria in broad terms: applicants had to be UK citizens between the ages of 21 and 40, to have been trained in any branch of science, to have the proven ability to learn a foreign language, and to be in a good state of health. Over 13,000 applications were received, a number that was gradually whittled down to four: Helen Sharman, Clive Smith, Major Timothy Mace, and Lieutenant-Commander Gordon Brooks. The names were announced in November 1989, initiating an intense period of media scrutiny that culminated in a live television programme hosted by Anne Diamond in which the 'winner' was announced. While the candidates were said to have felt uncomfortable at the level of media interest, they understood that 'the publicity would fuel the sponsorship', and that this was integral to the project as a whole.[26] At this point, the discourse around Project Juno diverged between the experiences of the prospective cosmonauts, for whom science and the prospect of spaceflight were the primary objectives, and the public narratives that surrounded them, in a mixture of orchestrated public relations and unscripted stories in the national media.

The media narratives around Project Juno were revealing in terms of the ways in which the final candidates were presented as potential icons of British exploration. Sharman recalled how one tabloid, *Today*, portrayed the other three finalists according to various male archetypes; 'the family man [...] the James Bond lookalike [...] the Officer and Gentleman', with Sharman reduced to the role of 'Token woman'.[27] When she revealed that she worked for the Mars Corporation, Sharman was predictably labelled 'the girl from Mars' in various news outlets, a moniker that assigned her as the alien outsider, while also infantilising her.[28] However, not all commentators drew consistently from this pool of sexist clichés, and, while initially suggesting that Sharman might contribute to Project Juno's 'showbiz image', the *Times* technology correspondent Nick Nuttall, in a series of articles, came to praise Sharman's 'considerable fortitude' and 'impressive cool head'.[29] Nuttall went on to portray in his articles a sense in which Sharman's mission had largely failed to capture the imagination of the British public and had not roused the sentiments of nationalism that might have been expected in an achievement of such significance:

> Even if the sponsorship promises had been realised, the nation was, in truth, never likely to have dusted down the Jubilee Union Jacks and huddled round the television sets as millions did when Apollo landed on the Moon.[30]

As Nuttall indicated, the funding model of Project Juno had broken down in the run-up to the launch, with sponsors failing to materialise in sufficient force. This resulted in the departure from the project of Heinz Wolff, whose involvement had been contingent on the possibility of conducting 'British scientific experiments' on *Mir*, and the mission was placed in considerable jeopardy before being rescued at the last minute by the Soviet-backed Moscow Narodny Bank.[31]

Here it is possible to identify several factors that contributed to the relative absence of nationalistic support from the UK media and general public as the mission was planned and executed: Firstly, a sense in which heroic British achievements were incompatible with a figure such as Sharman, a relatively normal, perhaps dispassionate, female figure from a non-military background, a feeling that was partly reflected through overt chauvinism in various media portrayals. British heroes of exploration had tended to conform to the image of rugged, adventurous men such as Ernest Shackleton or Edmund Hillary (even though the latter was a New Zealander), and Sharman simply did not fit into this prescribed role.[32] Second, the fact that it was not funded and directed by UK institutions perhaps naturally led to Project Juno having a fairly low national profile, a factor that was made more significant through figures such as Wolff withdrawing. Finally, the practices of private sponsorship lent Project Juno a somewhat superficial gloss, the selection procedure in particular being presented akin to a game show competition rather than as part of a national scientific endeavour of historic significance. Yet expressions of nationhood were not completely lacking in Project Juno, as became apparent during the launch itself and the mission aboard *Mir*.

When reading and listening to first-hand accounts of Helen Sharman's journey into space, a tension emerges between the formal geopolitical staging associated with her mission and her personal experience of spaceflight. As such, the geopolitical atmosphere near the end of the Cold War fed into the official choreography of the mission. The four Project Juno finalists arrived in the Soviet Union on the 12th November 1989, just two days after the fall of the Berlin Wall, while *Soyuz TM-12* launched in May 1991, just months before an attempted coup against Premier Mikhail Gorbachev brought about the rapid disintegration of the Soviet Union. Sharman and her understudy Tim Mace were insulated from the immediate effects of such events, and their role as cosmonauts-in-training during this period reflected an apparent ease in relations between East and West. Furthermore, during her stay on board *Mir*, Sharman participated in a telephone call with Gorbachev. Her account of this call spends more time explaining how it interrupted the crew's first hot meal in two days ('we felt it was worth breaking off our meal for that!') than recounting what was actually said.[33] Similarly, Sharman's last full day in the UK before the mission was 'dominated' by a meeting with Prime Minister Margaret Thatcher.[34] While it is hard to afford Project Juno any great geopolitical significance, it is perhaps best described as a symbolic reflection of a particular moment in UK-Russian relations, when the opening up of the Soviet regime was just about to reach a critical point. While these aspects of the mission were formal, they were also highly staged, artificial and performative in terms of their nationalistic intent. It is only when moving to the contrasting scale of personal experience that it is possible to gain deeper insights into Sharman's sense of individual agency as the first Briton in space.

Sharman's experience of being launched into space and spending seven days on *Mir* in many ways epitomised her hybrid identity as a UK citizen trained under the Soviet spaceflight regime – part astronaut, part cosmonaut. Her eighteen months of training at Star City not only involved physical, psychological and medical tests, but also a sense in which she was being trained in the traditions of Soviet spaceflight. This was apparent in the immediate run-up to the launch, when having 'learned to revere and cherish' the legacy of Yuri Gagarin, 'just as all the other cosmonauts did', Sharman participated in many of the traditions of Soviet cosmonauts, such as signing her name on the door of the accommodation block prior to her launch.[35] She took these performances in her stride, partly out of respect for her hosts and partly out of a genuine sense of reverence for the achievements of the Soviet space programme.[36] As part of the official framing of the mission, a series of symbolic representations also epitomised a sense of bilateral national partnership. This was apparent from the launch, with the *Soyuz* rocket emblazoned with both the Soviet and UK flags, as part of the official insignia of Project Juno, and in official photographs of the cosmonauts in front of both flags (Fig. 7.1). Unlike the overtly nationalistic flagging of the 'stars and stripes' on the lunar surface two decades earlier,

Figure 7.1 Crew photo for Soyuz TM-12, featuring cosmonauts Artsebarsky, Sharman and Kirikalyov

Source Credit: spacefacts.de/Joachim Becker

this form of national symbolism serves as an example of the 'unwaved flag', a banal nationalism which the cosmonauts, both Soviet and British, seemed to have tolerated rather than celebrated.[37]

Indeed, there is a sense in which national flagging was understood by Sharman as a routine duty rather than as an essential part of her role as the first Briton in space. By contrast, her overriding memories from her time on board *Mir* seem to have coalesced around two kinds of experience. Firstly, along with a strong sense of comradeship felt towards her crewmates, *Mir* itself took on a set of benevolent characteristics:

> The space station was my home and my feelings about it were more complete than I have ever felt for any flat or house I've lived in [...] the station had life![38]

The second overriding set of memories that Sharman recalled following the mission was associated with looking back at the Earth. Spending 'long periods at the window', Sharman explained how 'the world acquired an immensity that until then had only been academic'.[39] This sense of sheer size became apparent as the orbital pathway of *Mir* sent it rapidly across the surface of the Earth, covering large swathes of the planet at an altitude of around 200 miles. While marvelling at the beauty and immensity of the Earth, Sharman was also able to identify particular places on the surface:

> I could see the distinctive shape of Britain (often covered in cloud). I always had a warm and sentimental feeling whenever I could see parts of the world that I knew well[40]

Here, a sense of affective national identity is apparent, as Sharman depicts the familiar, homely feeling associated with Britain, which is also described as having an appearance of 'brown soil with bubbly greyer bits, which must have been the towns. Very uninspiring!'.[41] The mundane or everyday aspects of the image of Britain from space – the weather, the brown soil – contributed to Sharman's personal sense of national identity, but did not leech into expressions of nationalism. As she departed the space station at the end of her mission, Sharman chatted over the radio with the cosmonauts who had remained behind, and having left them a mix-tape of her favourite songs as a parting gift, the track 'World Outside Your Window' by the British folk singer Tanita Tikaram was relayed over the radio to the departing capsule.

While perhaps not being elevated to the status of national icon upon her return to Earth, Sharman spent several years touring the UK talking about her spaceflight experience, before deciding to withdraw from public life and continue her research career in chemistry at Imperial College London. One of the most prominent ways in which Sharman's mission to *Mir* has been remembered in the public realm is through the display of artefacts at the UK's National Space Centre in Leicester, which opened in 2001. Here, in

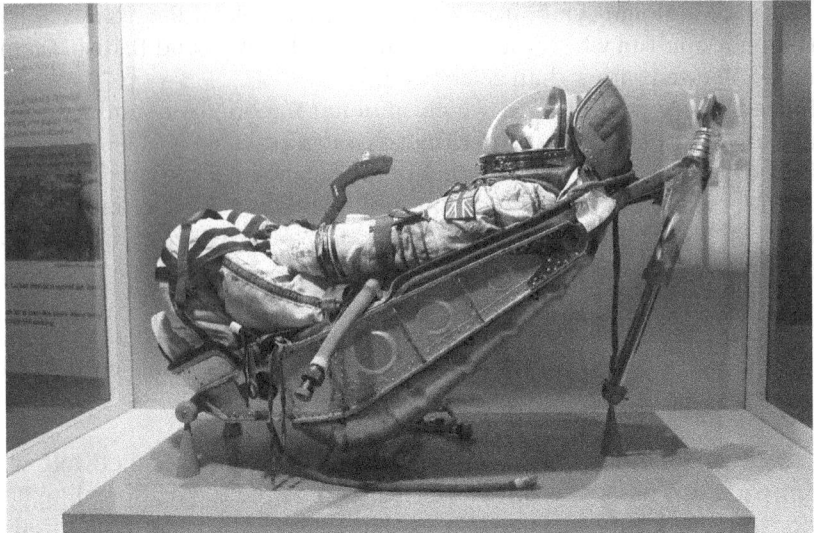

Figure 7.2 Helen Sharman's launch couch and spacesuit, exhibited at the National
　　　　　 Space Centre, Leicester

Source Credit: National Space Centre

the 'Into Space' gallery, the *Kazbek-U* Shock Absorbing Couch that carried Sharman into space and back to Earth is displayed, along with her
Sokol KV-2 Rescue Spacesuit, that she used in training (Fig. 7.2). Her light-
weight Cosmonaut Flightsuit and *Mir* Hygiene Kit (containing hairbrush,
toothbrush, toothpicks, comb and nail file) are also displayed. The National
Space Centre object page for the *Kazbek* couch describes the combined
couch-spacesuit display as intended 'to demonstrate the conditions for cosmonauts during launch and landing'.[42] Indeed, the exhibit is notable for the
way in which the human form is positioned, cramped and recumbent, the
only visible means of agency being the blue stick protruding between the legs,
which holds a button that was used to activate the ground-control communications system. Sharman herself is represented by a white mannequin within
the spacesuit.

This display to some extent re-enforces certain media narratives prior
to Sharman's mission, as she is presented as a passive traveller aboard the
Soyuz spacecraft, or 'more of a tourist than a cosmonaut', rather than being
actively heroic in the mould of the popularly imagined, male, adventurous
British explorer.[43] At the National Space Centre, the national flagging on
the spacesuit along with Sharman's prominent status as the first Briton in
space contrasts with the somewhat low-key status of the display. Indeed,
Sharman's public legacy was even said to have been 'written out of history'
in one account some years later, contributing to a sense in which Sharman's

achievements became relegated to a blind-spot in the national past.[44] Moreover, the nationalistic aspects to the story of the first Briton in space can be understood in terms of the personal sense of national identity that Sharman felt while orbiting the Earth in *Mir*, the banal forms of national flagging that decorated her mission in official representations, and the geo-political symbolism that surrounded the mission as a whole. What remains is an acknowledgement of what was *not* messaged in accounts of Sharman's experience in media stories, and perhaps even a sense of forgetting in the public memorialisation of the event. This acts as a reminder of the ways in which national memory is socially, culturally and politically arbitrated, and reinforces a sense in which the anticipation of British achievement in space that had been building for much of the twentieth century, had ultimately been left unresolved in the national psyche.

Beagle 2: public memory and national popular culture

The next time that British involvement in space exploration caught the public's attention, it involved a strong sense of nationalistic flagging by the mission's protagonists. 'Beagle 2', a Mars lander carrying instruments to detect signs of life, crashed onto the Martian surface on Christmas Day 2003, instantly becoming a national news story of plucky British failure. The project was conceived by the British space scientist Colin Pillinger, a specialist in the analysis of extra-terrestrial materials, and one of the first scientists outside of the USA to examine samples of Moon rock brought back by the Apollo missions.[45] It was, however, the search for extra-terrestrial life that had inspired the concept of Beagle 2, as imagined narratives of life on Mars that were stirred so effectively by H G Wells a century earlier (see Chapter 2) were re-enacted in a captivating piece of scientific theatre. In 1996, NASA announced the publication of a research paper claiming that a meteorite of Martian origin that had landed in Antarctica in 1984 contained ancient signs of microbial life.[46] This 'discovery' became a major media event, with the US President Bill Clinton stating that 'its implications are as far-reaching and awe-inspiring as can be imagined', in a statement that captured the interface between science, the sublime and the imaginative.[47] Following the NASA announcement, Pillinger's comment piece in *Nature*, along with Open University colleagues Monica Grady and Ian Wright, provocatively engaged the scientific community with the notion of life on Mars.[48] The metaphorical 'can of worms' referred to in the paper's title implicitly acknowledged the form of the supposed fossilised microscopic bacteria found on the Martian meteorite, morphology that was seen as inconclusive from the outset by other scientists in the field.[49] Pillinger's ensuing idea, and what became the sole focus of the rest of his career, was to secure funding for, design and project-manage a British Mars lander that would look for signs of life on the Red Planet.

British involvement in space science at this time had been dependent on securing payload space aboard European and American rockets, with UK agencies eligible to tender for inclusion in the scientific missions of the European Space Agency (ESA). Seeking to take advantage of the anticipated opposition of Mars in 2003, whereby Mars and Earth would be relatively close to one another, ESA started working on the Mars Express mission in 1997. Its primary objective was to establish a Mars-orbiting spacecraft that would analyse the Red Planet remotely, with the possibility of also deploying a lander to the surface. Beagle 2 was proposed to fulfil this secondary objective, to follow in a history of Mars landers that started in the 1970s with the Soviet Union's Mars probes and NASA's Viking programme, and continued with NASA's successful Pathfinder probe in 1997.[50] In the midst of Beagle 2's uncertain development, Pillinger led a variety of fundraising efforts focussed on UK government science programmes as well as private finance initiatives. In doing this, Pillinger co-operated closely with national media outlets, also working with the advertising giant M&C Saatchi. As one commentator put it, 'he assiduously pressed all the old cultural buttons' as part of his efforts to enrol the British public in Beagle 2's anticipated success.[51]

Pillinger's attitude to promoting Beagle 2 was connected inherently to his own personal ethos of science communication, by which he understood that 'if you want the public to take an interest, then you have to let them join in the excitement'.[52] Here, the feeling of excitement became central to the popular affective nationalism through which Beagle 2 was imagined and promoted, spliced with nostalgia for Britain's imagined past. Hence, Pillinger corralled an imagined nationwide community through a series of media engagements and cultural endorsements in his attempts to shape the public narrative of Beagle 2, with the steadfast support of his wife Judith. Perhaps the foremost of these strategies was the naming of the lander itself, which was one of Judith Pillinger's ideas. Reacting to initial suggestions for names during a journey back home from an ESA meeting in Paris, Colin quotes Judith as saying;

If it's going to be British then it ought to be Darwin, not Pasteur.[53]

The chosen name 'Beagle 2' implied a genealogy that aligned the space probe with Charles Darwin's 1831–36 voyage aboard the HMS *Beagle*, from which he amassed evidence to support the theory of evolution by natural selection. Indeed, Pillinger hoped that Beagle 2, by finding evidence of life on Mars, would propel evolutionary biology into a new era. At the same time, the interplanetary journey of Beagle 2 would mimic the historic voyage of HMS *Beagle*, drawing upon the symbolic interface between outer space and the seas that has long characterised narratives of space exploration. All things considered, it was clear that the story of Beagle 2, and its wider cultural and political framing in accounts such as Pillinger's *Beagle 2 Diaries*, would come to define the mission. Such frameworks have been common in maritime history, with records such as official hydrographical reports inscribing their subjects as 'products of national and imperial enterprise' in narratives of exploration and adventure, traits that can also be identified in the narration of Beagle 2.[54]

The invocation of HMS *Beagle* brought to mind a time in which British imperial networks and associated global naval dominance, particularly over the French, fostered scientific developments and enabled new discoveries in the natural realm. While enrolling the expansionist language of global exploration, Beagle 2 also invoked nostalgia for the scientific achievements of the imperial age. To this effect, comparative features of HMS *Beagle* and Beagle 2 were put on public display in an exhibition at London's National Maritime Museum in December 2002, with an associated book also released.[55] The exhibition showcased artefacts from Darwin's voyage, including biological specimens and scientific instruments, alongside animations of Beagle 2's journey to Mars, and its design plans. The accompanying press release quoted the museum's director Roy Clare reflecting on 'the tradition of heroic exploration, once by sea and now manifest in space'.[56] Further comparisons were made between the two Beagles, from the careful allocation of space on board the vessels, to the types of instrument designed to measure geochemical, atmospheric and mineralogical qualities at their corresponding destinations. Both were said to represent the 'state-of-the-art in exploration for their respective generations, leading the way in science and technology, to say nothing of adventure', hinting at a broader appeal beyond the purely scientific.[57]

The narration of Beagle 2 sought to trace the lineage not just of Darwin's HMS *Beagle*, but also the several other HMS *Beagle* vessels that have been named by the Royal Navy, from the first in 1766 to the ninth in 1967. Together, these ships established a naval heritage that spanned conflicts including the Napoleonic Wars, the Crimean War and the Second World War, during which the eighth HMS *Beagle* patrolled the Dover straits, took part in the invasion of North Africa, and 'was involved in freeing the Channel Islands'.[58] It is clear that, along with establishing a genealogy of scientific exploration associated with the name Beagle, Pillinger was seeking to register stories of British military triumph, assigning Beagle 2 as the plucky underdog, the adventurous explorer, and as harbinger of a glorious victory. This attitude of British pre-eminence is encapsulated by Pillinger's comment that, even though he 'counted thirteen countries contributing' to the endeavour, and notwithstanding its complete reliance on the ESA launcher and Mars Express, Beagle 2 'is a quintessentially British project'.[59]

As well as looking to past triumphs of British science and exploration to shape the narrative of Beagle 2, Pillinger and his team sought to engage with contemporary popular culture icons that were associated with the emerging 'Britpop' oeuvre. By the late 1990s, a resurgence of interest in British pop music, that itself referenced the cultural heyday of the 1960s, emerged alongside the 'Young British Artists' group including Tracey Emin and Damien Hirst, who headlined the renowned 'Sensation' exhibition at the Royal

Academy of Arts in 1997. Combined with the landslide election of the 'New Labour' government under Tony Blair, there was a sense in which politics, art and culture were engaging confidently with aspects of Britishness in a way that had not been witnessed in recent history.[60] Looking more carefully, however, some critics have commented on the ways in which British national identity in this period was in fact fragmenting, highlighting the smaller-scale nationalisms that fed into independence and devolution movements among the different nations of the United Kingdom.[61] As such, any re-emergence of 'Britishness' in national culture could be seen as Englishness in disguise, having more in common with English cultural traditions than with any unifying concept of British national identity.[62] There is also a sense in which, in an era of cultural fragmentation, movements like Britpop represented 'a nostalgic turn to the past – as a harking back to a Britain that has been lost', an ultimately superficial engagement with national culture.[63]

Nevertheless, the zeitgeist of the era was manifested in the actual design of Beagle 2, with artwork by Damien Hirst and a musical composition by the rock band Blur incorporated into the lander's hardware and software. These connections were established when Blur's drummer Dave Rowntree and bassist Alex James contacted Pillinger with the idea to incorporate a short musical refrain into Beagle 2's digital call sign, to be transmitted once it had successfully landed on the Martian surface. Rowntree commented at the time that 'Martian bacteria love Britpop', encapsulating the upbeat irreverence of the era, or perhaps its meaninglessness.[64] Rowntree and James later introduced Pillinger to Damien Hirst, who in consultation with Beagle 2 scientists, created a special version of one of his well-known 'spot paintings' to be incorporated into the spacecraft's structure. According to Pillinger, the artwork 'fitted all the criteria, while fulfilling an essential scientific need', which was to help calibrate on-board spectroscopic instruments following the spacecraft's turbulent descent to the Martian surface.[65] It consisted of a selection of pigments set into a space-qualified aluminium framework, including hues of yellow and red that were expected to match the colours of the Martian environment, as well as white, black, blue and green, the latter representing Planet Earth. In a reciprocal act, both Blur and Damien Hirst produced independent works named 'Beagle 2' – the former a B-side track from the 1999 single 'No Distance Left to Run', the latter another variation of the artist's spot paintings. In these designs, and the publicity that came with them, Beagle 2 had become fully incorporated into millennium-era British popular culture.

It might be assumed that Blur and Hirst's compositions for Beagle 2 acted primarily as technical elements in the lander's series of operational procedures. However, the calibration target notably was 'explicitly staged as a piece of art', including in an exhibition at London's White Cube gallery prior to the launch, while Blur's digital call sign has been described as no more than 'a marketing strategy to attract the attention of mass media and thus possible public and private sponsorship'.[66] Art and music were not necessary to the mission's technical operations, but were important in getting Beagle 2 off the ground by using cultural capital as part of the appeal to public and private funding bodies. For one critic, this relationship 'exposes a sincere lack of metaphysical beliefs beyond a capitalist consumerism which recycles and re-exploits historical fetishes and icons'.[67] Here, the science of space exploration is seen to have incorporated creative products as part of a consumerist agenda, resonating with some critiques of millennium-era popular culture as being superficial in nature and constitutive of a postmodern re-hashing of past cultural identifiers. This feeling was captured in a satirical cartoon by Andrew Birch as part of his 'Young British Artists' series, that imagined Martian art critics reacting to the Beagle 2 lander (Fig. 7.3).

Beagle 2's mission culminated in its landing on Christmas Day 2003. As Pillinger recounted, 'whilst homes all over Britain were listening to Her Majesty, we were straining electronic ears for signs of a missing dog'.[68] In this way, a further attachment to the British national psyche was established with the alignment of the landing event with the Queen's annual televised address to the nation. When the call sign failed to materialise, doubts began to surface as to the mission's success, and after several further opportunities to detect Beagle 2 in early 2004 had also passed, a joint UK-ESA commission of inquiry was set up to establish what went wrong. The report uncovered a series of management and technical shortcomings, with inadequate safeguarding and testing measures likely to have resulted in the failure of some of Beagle 2's landing mechanisms, potentially the airbags, the parachute, or electronic components. Moreover, the report stated that 'Mars

Figure 7.3 Andrew Birch 'Young British Artists' cartoon from 2002 featuring the Beagle 2 lander

Source Credit: *Private Eye* and Andrew Birch

Express and Beagle 2 should have been managed as an integrated mission', rather than allowing Beagle 2 to proceed with its 'unconventional' management arrangements.[69] Notwithstanding these valid criticisms, conventionality certainly wasn't a prime characteristic of Beagle 2, and perhaps the spacecraft would not have reached Mars at all without some of the maverick qualities of Pillinger himself.

It is fair to say that the relationship between ESA and Beagle 2 came to represent a broader set of cultural and political tensions between the UK and continental Europe. Indeed, there is a sense in Pillinger's writings that 'Europe', or at least the European Space Agency, was an interfering presence, wanting to 'raise as many obstacles to Beagle 2 as they could', and his post-hoc accounts certainly involve an element of righteous score-settling.[70] Pillinger's strategy seems to have been to overcome what he saw as stifling European bureaucracy through deploying the 'best of British' endeavour, from Darwin to Britpop. Here is a sense in which nationalism can emerge in a reactionary manner, conserving selected traditions of a nation in opposition to a form of internationalism that is seen as threatening and all-consuming. We can also understand Beagle 2 in association with what has been described as an 'aura of nationalism', incorporating broader affective resonances, cultural narratives and imagined national communities, not just flag-waving and pictorial symbolism.[71] Indeed, while Beagle 2 may have failed scientifically, it did succeed in raising the profile of British space science in the new millennium.

Tim Peake: performing nationalism in space

The first decade of the twenty-first century saw a gradual revival of interest in space exploration by the UK government, foregrounding the emergence of an officially sanctioned form of outer-space nationalism. With the establishment of the UK Space Agency in April 2010, a new strategy for space was initiated, focussed on the private space industry. This was spurred by the growth of companies in space manufacturing, space applications and space ancillary support services, which had contributed to the space sector's growth rate of 7.5% per annum by 2011.[72] The new UK Space Agency served mostly to amalgamate existing departments that dealt with space policy across the environmental, scientific and industrial sectors of government, while also providing the opportunity for a re-branding exercise. At the UKSA launch, science minister Lord Drayson described the new agency as 'a body with a firm grip on the future [with] the muscle to negotiate strongly on the UK's behalf, and to command the respect of both academia and industry', in a statement that conjured up some interesting imagery, but ultimately lacked any direct meaning.[73] The new logo was unveiled, a design that strongly incorporated the Union Flag (Fig. 7.4). It sought ostensibly to depict the UK as a dynamic, forward-thrusting power in space, but perhaps conversely gave a sense of the fragmented nature of the UK space industry.

Figure 7.4 Tim Peake at the launch of the UK Space Agency in April 2010

Source Credit: UK Space Agency

It also offered a contrasting visual language to other national space agency logos, the majority of which have shunned explicit national symbolism in their pictorial design, tending to favour instead stylised versions of the 'Apollonian gaze' from an imagined Earth orbit.[74] As such, the launch of the UK Space Agency served to convey a tension between the expansive discourse of national achievement in space, and the mundane realities of job creation and maintenance of science and engineering skills, in attempting to derive meaning from the UK's broader space endeavour.

Pressing the button to unveil the UK Space Agency logo at the launch event was Tim Peake, who was introduced by Drayson as 'the UK's first official astronaut'.[75] The label accentuated Peake's status as a government-sponsored member of the European Space Agency's Astronaut Corps, but it also served to de-legitimise the prior achievements of Helen Sharman as unofficial, with one commentator suggesting that Sharman was 'largely forgotten amidst the fanfare that surrounded Tim Peake'.[76] Certainly, during the 1980s and 1990s, successive UK governments had opted out of ESA's human spaceflight programme, as well as funding of the International Space Station, severely limiting the opportunities for prospective British astronauts. Furthermore, a number of 'British-born' NASA astronauts who had taken up US citizenship went into space in the 1990s and early 2000s, also falling into this 'unofficial' category of British astronauts.[77] Eventually though, a deal was struck by Drayson's successor as science minister, David Willetts, to provide a one-off payment of £16million to plug an ESA budget gap, in return for securing Tim Peake's place on a 2015 mission to the ISS.[78] Although this represented a greater willingness to be involved in

ESA activities, the UK was still only contributing a small fraction of ESA's budget in comparison to Italy, France and Germany. Indeed, the nationalities of the European Astronaut Corps are roughly in accordance with the funding structure of ESA, with Italian, German and French astronauts selected in greater numbers than other European nationals. As the only British representative of the European Astronaut Corps, Peake has performed the hybrid role of national icon and ESA astronaut, and one of the ways in which this duality became apparent is in the acknowledgement of his background in the British armed forces. Originally enlisted in the Army Air Corps, Peake served as a reconnaissance pilot, an instructor pilot, and a test pilot, progressing to the rank of Major. With ESA being a multi-national civilian organisation, Peake's military rank is downplayed in official profiles and press releases, in contrast to the UK media, in which he is commonly referred to as 'Major Tim Peake'. Here, the complexities of identity and national belonging in astronaut profiling are highlighted, as are their connections to political and financial imperatives in space policy.

Despite Helen Sharman's mission to *Mir* occurring some twenty-four years earlier, Tim Peake's final preparations and launch to the ISS from Baikonur in Kazakhstan were remarkably similar, as he participated in the same Soviet-originated rituals in Star City and was launched in an almost identical *Soyuz* rocket. His team-mates on board *Soyuz TMA-19M* were US astronaut Tim Kopra and Russian cosmonaut Yuri Malenchenko, who performed a manual docking procedure with the ISS, the automated system having failed on first approach. However, unlike Sharman, Peake's space-flight experience was well-supported in the national media. Over the duration of his six-month ISS mission, he made live or recorded appearances from space at almost every major UK cultural or sporting event, from the Edinburgh Hogmanay celebrations to the Six Nations rugby tournament. Peake also made use of social media throughout his training and mission duration, updating blogs, posting photos and tweets to connect to a broad public audience. Notwithstanding this substantial media schedule, Peake's activities on the ISS primarily consisted of space-station maintenance, conducting scientific experiments, and performing educational outreach activities. Peake later commented that life on board the ISS 'quickly falls into a routine and becomes surprisingly normal', and that this 'normalisation' becomes 'an essential process' in safely and effectively operating in space.[79] The normality of spaceflight here becomes a characteristic component of the nationalist messaging that occurred during Peake's mission, and two performances demonstrate this in contrasting ways: his spacewalk on 15th January 2016 and his running of the London Marathon on 24th April 2016.

Part of the maintenance of the International Space Station involves astronauts conducting spacewalks, or Extra-Vehicular Activities (EVAs), around the outside of the station to complete a variety of technical tasks. The principal objective of Tim Peake's spacewalk, conducted in tandem with Tim Kopra, was to replace a faulty voltage regulator at the base of one of the space station's solar power arrays. This part of the EVA had to occur during one of the station's 45-minute 'night' phases of orbit, when the arrays were not taking in energy from the Sun. Peake additionally was tasked with laying a 28-metre-long communications cable, before completing his spacewalk within a planned duration of over six hours.[80] In the event, the spacewalk concluded in under five hours, the main objectives having been achieved, due to a bubble of liquid that was identified inside the helmet of Kopra's spacesuit, that triggered the mission's precautionary termination. The routine functionality of the EVA was interrupted in the first few minutes by an acknowledgement of Tim Peake being the first British astronaut to embark on a spacewalk, wearing the Union Flag on his spacesuit:

> Reid Weisman (ground control): 'Hey Tim, it's really cool seeing that Union Jack go outside, since it's explored all over the world, and now it's explored space.'
> Tim Peake: '[laughter] … It's great to be wearing it. It's a huge privilege – a proud moment.'[81]

Here, the audio recording relays the sincerity with which Peake made his comments and his clear pride at wearing the national flag in space. The comments seem unscripted, with nationalistic themes introduced by the ground control operator and spontaneously affirmed by Peake as he emerged from the airlock. The exchange centres on the presence of a material object in space – the Union Flag patch attached to the shoulder of the spacesuit – and the agency afforded to it by Peake. Prior research on NASA's Space Shuttle missions has demonstrated how the iconography of space mission patches has continued to reflect and reify nationalistic sentiment, a process that is further encountered here in the more direct symbolism of national flagging in space.[82] In this case, however, it is the materiality of the flag patch, in combination with its embodiment by Tim Peake, that generates the potent national symbolism of the event as it occurs, rather than in the post-hoc environment of national museum spaces and space memorabilia.

The contrast between the routine normalities of space-station maintenance and the enactment of a historic 'first' for British spaceflight is a striking theme of Tim Peake's spacewalk. He later reported that he 'went to sleep that night a very proud Briton – with the Union Flag from my spacesuit Velcroed to the wall in my crew quarters'.[83] It was clear that national pride had a personal affective resonance with Peake, while the video footage of his spacewalk was broadcast by television and online news outlets in the UK and around the world. It is perhaps helpful to think of Peake's spacewalk as a performance through which elements of nationalistic meaning can be derived, mindful of work in cultural geography that acknowledges the ways in which bodies move affectively, culturally and politically.[84] Each astronaut movement or 'translation' across the surface of the ISS is choreographed carefully, while some are rehearsed in advance, similar to a dance performance. The spacewalk is filmed for specific audiences, consisting of not only the ground-control operations team, but also a national and world-wide viewership, from a variety of cameras stationed on the exterior of the ISS as well as on the helmet of each astronaut. Furthermore, the mission's narrative is punctuated with dramatic incidents, such as Peake's emotional 'proud moment' and the possibility of danger in the mission's premature termination. It is therefore apparent that, as well as serving an essential function in space-station maintenance, Tim Peake's spacewalk acted as a process through which nationalism continued to establish its presence, through the combined vectors of performance, affect and material culture.

A further iteration of these themes occurred as Peake remotely completed the London Marathon in space, some four months into his mission on board the ISS. While astronauts and cosmonauts undertake daily physical exercise using resistance machines in the ISS gym, in order to counteract the muscle and bone wastage that occurs in space, the zero-gravity environment presents some unique challenges for running. This can only be achieved through the use of a tethered harness that pulls the wearer into a treadmill, thereby enabling sufficient traction. Peake was able to use his personal leisure time to prepare for the run, while he also recorded a video to officially start the London Marathon. In his running gear, positioned in front of his spacesuit with Union Flag patch, Peake's message culminated in a count-down to set the runners off:

> Hi, I'm ESA astronaut Tim Peake, on board the International Space Station. It's a huge honour to be asked to be the official starter of the 2016 London Marathon. I'm really excited to be able to join the runners on Earth, from right here on board the space station. Good luck to everybody running, and I hope to see you all at the finish line. 10 … 9 … 8 … 7 … 6 … 5 … 4 … 3 … 2 … 1 … go![85]

Meanwhile, Peake started the marathon somewhere above the Pacific Ocean and finished in a time of three hours, thirty-five minutes.[86] The London Marathon is known worldwide for its route along many of the iconic land-marks of the UK capital, and its starting point in Greenwich is home to historic national centres of maritime and astronomical science, including the Royal Observatory and Royal Naval College. Indeed, the ISS itself operates on Greenwich Mean Time, despite the fact that its mission control centre is located in Houston, Texas, and it passes over all the time zones of Earth in each orbital period of ninety minutes. As Peake started the London Marathon, he reinforced the significance of Greenwich as an international symbol of timekeeping, and London as a global city, in a virtual act in unison with the thousands of runners in London.

This was not the first time that Tim Peake had run the London Marathon, having previously completed it in 1999, and his abiding memory of that event was 'the overwhelming atmosphere of camaraderie, fun and support from the crowd throughout the entire course'.[87] While completing the marathon in space, Peake was able to view footage of the runners on the ground through a live video link. He later commented that he drew 'encouragement from the crowds and the thousands of other runners', as some element of the event's atmosphere was translated to the International Space Station.[88] This, when considered in the context of the London Marathon's iconic representation of the nation, resonates with arguments about the ways in which nationalism can be understood 'as an atmosphere'.[89] Such work has explained how large sporting events, such as the London 2012 Olympic Games, have served to process 'the nation's affective, emotional and atmospheric resonances' through communal participation in the 'atmosphere' of the event itself.[90] Peake's marathon demonstrates how the affective atmosphere of nationalism can even be sensed in outer space, more specifically in the low Earth orbit that took Peake around the globe twice over during his marathon run.

Upon returning to Earth in the *Soyuz* descent module on the 18th June 2016, along with his two crewmates, Tim Peake began the process of rehabilitation that is necessary after a six-month duration in the weightless environment of the ISS. Having been transported from the landing site on the plains of Kazakhstan to the ESA headquarters in Cologne, Germany, Peake conducted a press conference and participated in his first interview, just one day after his landing, describing the feeling of being back on Earth as 'something akin to the world's worst hangover'.[91] Commitments such as these reflected Peake's role as a continuing member of the ESA Astronaut Corps, and he continued to act as the research subject in a variety of experiments into the effects of spaceflight on human physiology. Over the next few months and years, Peake featured in a variety of national media, from an interview in *Hello* magazine to an appearance at the 2017 UK Space Conference, an event organised by the UK Space Agency biennially since 2011, maintaining his profile as the UK's sole astronaut on active duty.

Perhaps the most effective way in which the story of Tim Peake's mission to the ISS has been told to a British audience is through the recent touring exhibition of the *Soyuz TMA-19M* descent module and *Sokol KV-2* emergency spacesuit in which he returned to Earth. The Science Museum Group estimates that this exhibition has been seen by 1.3 million members of the public throughout its 20-month duration.[92] Starting in September 2017 at the National Science and Media Museum in Bradford with just the *Soyuz* module, it continued through Shildon in County Durham, York, Manchester, Edinburgh, Peterborough, Cardiff and Belfast, thereby taking place in all the nations of the United Kingdom, before finally being put on permanent display in London's Science Museum in 2019. The capsule's charred appearance reveals the story of its violent descent from the International Space Station, while visitors are able to see inside the capsule to view the shock-absorbing couches, instrument panel and survival equipment, and gain a sense of the close atmosphere within. Peake's *Sokol* spacesuit was included in the latter part of the tour and went on permanent display at the National

Space Centre in Leicester in 2019. The *Sokol* suits are tailored to each wearer, designed for the recumbent position in the shock-absorbing couch, leading to a distinct 'cosmonaut stoop' when worn upright.[93] While Helen Sharman's *Sokol* suit was displayed within its couch, as described above, Tim Peake's *Sokol* suit is exhibited in the standing position, giving it a greater sense of agency and purpose. This reflects Peake's elevated 'official' status in British spaceflight, and is indicative of the clear nationalistic messaging that was attached to his spaceflight experience. It is perhaps easy to forget that none of Peake's public-facing work during and around his mission, or its cultural and political framing, was inevitable or accidental. When considering US astronaut Scott Kelly's 'year in space' that overlapped with Peake's mission to the ISS, themes of endurance and hardship are foregrounded in his memoirs, while Kelly's comradeship with the Russian ISS cosmonauts is said to have 'provided some hope in dire geopolitical times'.[94] What marked out Peake's experience were decades of pent-up anticipation of the British exploration of space, that, although always contingent on a range of international partnerships, was finally released into the public realm.

Conclusion: nationalism and British cultures of space exploration

Around the turn of the twenty-first century, a series of British endeavours in space exploration were connected to expressions of nationalism through a diverse interplay between bodies, objects and spaces. In Helen Sharman, the first Briton in space was celebrated in an underwhelming manner, the connections with the Soviet space programme and her status as an unconventional, female British icon of exploration feeding into this reception, despite her positive experience of spaceflight. By contrast, a less-successful space mission, in Beagle 2, was commonly characterised in relation to a multitude of symbols and icons of nationalism, awkwardly marrying the two cultural touchstones of historic imperial exploration and millennium-era pop culture. Tim Peake's mission to the International Space Station engendered more conventional or official representations of nationalism, that were supplemented by his performances in space. These examples in the first instance demonstrate the reach of nationalism, its prevalence beyond the bounds of the nation-state, and the importance of outer space as a vector of nationalism to nation-states other than the competing powers of the Cold War era of spaceflight. This has been seen not only in terms of specific experiences in space, but importantly, in the national stories that are told about outer space on the ground and across the airwaves in Britain, including in print and broadcast media, scientific research programmes and national cultural events. Given the substantial cultural and political accumulation of expectation that has been highlighted throughout this book, from science-fictional visions to detailed plans for future space missions, it is perhaps unsurprising that British achievement in space would bring with it a wide range of nationalistic signifiers, and that these expressions would come to supersede allusions to internationalism that emerged briefly in the mid-twentieth century. However, across the three examples highlighted in this chapter, it has become apparent that memory, gender and culture also came to define nationalism in outer space, highlighting how nationalism is as much about forgetting as it is about recalling the achievements of the national past. Hence, while nationalism has become a central part of the geographies of outer space in Britain, it has to be understood as a nuanced and partial phenomenon that will continue to shape British activities in space in the years to come.

Notes

1 John M Logsdon, 'Space in the Post-Cold War Environment', in *Societal Impact of Spaceflight*, ed. by Steven J Dick and Roger D Launius (Washington DC: NASA, 2007), 89–102.
2 Roger D Launius, 'What Are Turning Points in History, and What Were They for the Space Age?', in *Societal Impact of Spaceflight*, ed. by Roger D Launius and Steven J Dick (Washington DC: NASA, 2007), 19–40.
3 John Agnew, 'Nationalism', in *A Companion to Cultural Geography*, ed. by James Duncan, Nuala Johnson and Richard Schein (Oxford: Blackwell, 2008), 223–237 (p.225).
4 Benedict Anderson, *Imagined Communities – Reflections on the Origins and Spread of Nationalism* (London: Verso, 1983); Agnew, 'Nationalism'; Nuala Johnson, 'Cast in Stone: Monuments, Geography and Nationalism', *Environment and Planning D: Society and Space,* 13 (1995), 51–65.
5 Michael Billig, *Banal Nationalism* (London: SAGE, 1995), p.6.
6 Tim Edensor, *National Identity, Popular Culture and Everyday Life* (Oxford: Berg, 2002); Rhys Jones, 'Relocating Nationalism: On the Geographies of Reproducing Nations', *Transactions of the Institute of British Geographers,* 33 (2008), 319–334.
7 Daniel Sage, *How Outer Space Made America: Geography, Organization and the Cosmic Sublime* (Farnham: Ashgate, 2014).
8 De Witt Douglas Kilgore, *Astrofuturism: Science, Race and Visions of Utopia in Space* (Philadelphia: University of Pennsylvania Press, 2003); Howard E McCurdy, *Space and the American Imagination* (Washington DC: Smithsonian Institution Press, 1997); James Spiller, *Frontiers for the American Century: Outer Space, Antarctica and Cold War Nationalism* (New York: Palgrave Macmillan, 2015).
9 Elisabeth Millitz and Carolin Schurr, 'Affective Nationalism: Banalities of Belonging in Azerbaijan', *Political Geography,* 54 (2016), 54–63 (p.61).
10 Angharad Closs Stephens, 'The Affective Atmospheres of Nationalism', *cultural geographies,* 23 (2016), 181–98; Peter Merriman, and Rhys Jones, 'Nations, Materialities and Affects', *Progress in Human Geography,* 41 (2017), 600–617.
11 Andrew Maclaren, in Oliver Dunnett and others, 'Geographies of Outer Space: Progress and New Opportunities', *Progress in Human Geography*, 43 (2019), 314–336.
12 Maclaren, 'Geographies of Outer Space' (p.333).
13 Tamar Mayer, 'Embodied Nationalisms', in *Mapping Women, Making Politics – Feminist Perspectives on Political Geography*, ed. by Lynn A Staeheli, Eleonore Kofman and Linda J Peake (London: Routledge, 2004) 153–168 (p.153).

14　Asif Siddiqi, *Challenge to Apollo: The Soviet Union and the Space Race, 1945–1974* (Washington DC: NASA, 2000), p.295.

15　James T Andrews, 'In Search of a Red Cosmos: Space Exploration, Public Culture, and Soviet Society', in *Societal Impact of Spaceflight*, ed. by Roger D Launius and Steven J Dick (Washington DC: NASA, 2007), 41–52 (p.41).

16　Daniel Sage, 'Giant Leaps and Forgotten Steps: NASA and the Performance of Gender', in *Space Travel and Culture: From Apollo to Space Tourism*, ed. by Martin Parker and David Bell (Oxford: Blackwell, 2009), 146–163 (p.153).

17　Tom Wolfe, *The Right Stuff* (New York: Bantam, 1979).

18　Fraser MacDonald, 'Geopolitics and "the Vision Thing": Regarding Britain and America's First Nuclear Missile', *Transactions of the Institute of British Geographers,* 31 (2006), 53–71.

19　Fraser MacDonald, 'Space and the Atom: On the Popular Geopolitics of Cold War Rocketry', *Geopolitics,* 13 (2008), 611–634.

20　Jonathan Amos, 'When Britain had a small astronaut corps', *BBC News* (2010) http://www.bbc.co.uk/blogs/thereporters/jonathanamos/2010/03/when-britain-had-a-small-astro.shtml [11th August 2020].

21　Helen Sharman and Christopher Priest, *Seize the Moment – the Autobiography of Helen Sharman* (London: Victor Gollancz, 1993), p.51.

22　David Shayler and Ian Moule, *Women in Space – Following Valentina* (Berlin: Springer Praxis, 2005), p.314.

23　Sharman and Priest, *Seize the Moment.*

24　Bettyann Holzmann Kevles, *Almost Heaven – the Story of Women in Space* (Cambridge, MA: MIT Press, 2006), p.136.

25　Jamie Grierson, 'Heinz Wolff, scientist and Great Egg Race presenter, dies at 89', *Guardian* (2017) https://www.theguardian.com/science/2017/dec/16/heinz-wolff-scientist-and-great-egg-race-presenter-dies-at-89 [11th August 2020].

26　Sharman and Priest, *Seize the Moment* (p.75, p.79).

27　Sharman and Priest, *Seize the Moment* (p.92).

28　'The Girl from Mars is to be UK's first in space', *Liverpool Echo*, 22nd February 1991, p.2.

29　'Day of reckoning for first British astronaut', *The Times*, 22nd February 1991; 'Mission circus gives way to countdown', *The Times*, 11th May 1991, p.8.

30　'Mission circus gives way to countdown', p.8.

31　'Heinz Wolff turns back on Soviet space hope', *The Times*, 15th January 1991, p.2.

32　Peter Hansen, 'Coronation Everest: The Empire and Commonwealth in the 'Second Elizabethan Age', in *British Culture and the End of Empire*, ed. by Stuart Ward (Manchester: Manchester University Press, 2001), 57–72.

33　Sharman and Priest, *Seize the Moment* (p.101).

34　Sharman and Priest, *Seize the Moment* (p.140).

35　Sharman and Priest, *Seize the Moment* (p.27).

36　Although the Soviet Union had a long tradition of female cosmonauts, Sharman was still treated differently because she was a woman. As part of what was intended as a light-hearted prank, she was made to smuggle on board *Mir* a pink chiffon dress, which she wore 'to dinner' on one occasion in the space station.

37　Billig, *Banal Nationalism* (p.93).

38　Sharman and Priest, *Seize the Moment* (p.136).

39　Sharman and Priest, *Seize the Moment* (p.132).

40　Sharman and Priest, *Seize the Moment* (p.131).

41　Sharman and Priest, *Seize the Moment* (p.132).

42　National Space Centre 'Kazbek-U Shock-Absorbing Couch', *National Space Centre Collections Online* (2019): http://collections.spacecentre.co.uk/object-1999-3 [1st March 2019].

43 Holzmann Kevles, *Almost Heaven* (p.159).
44 Colin Drury, 'Blast off! Why has astronaut Helen Sharman been written out of history?', *Guardian* (2016) https://www.theguardian.com/lifeandstyle/2016/apr/18/blast-off-why-has-astronaut-helen-sharman-been-written-out-of-history [1st March 2019].
45 John Zarnecki, 'Pillinger, Colin Trevor, 1943–2014', *Oxford Dictionary of National Biography* (2018) https://doi.org/10.1093/odnb/9780198614128.013.108749 [11th August 2020].
46 David S McKay, 'Search for Past Life on Mars. Possible Relic Biogenic Activity in Martian Meteorite ALH84001', *Science*, 273 (1996) 924–930.
47 Bill Clinton, 'President Clinton Statement Regarding Mars Meteorite Discovery', NASA / Jet Propulsion Laboratory (1996) https://www2.jpl.nasa.gov/snc/clinton.html [8th April 2019].
48 Monica Grady, Ian Wright and Colin Pillinger 'Opening a Martian can of worms?', *Nature* 382 (1996) 575–576.
49 Juan-Manuel Garcia-Ruiz, 'Morphological behavior of inorganic precipitation systems', *Proc. SPIE* 3755 Instruments, Methods, and Missions for Astrobiology II (1999).
50 Jason Dittmer, 'Colonialism and Place Creation in "Mars Pathfinder" Media Coverage', *The Geographical Review,* 97 (2007), 112–130.
51 Francis Spufford, *Backroom Boys – the Secret Return of the British Boffin* (London: Faber and Faber, 2003), p.223.
52 Colin Pillinger, *My Life on Mars – the Beagle 2 Diaries* (London: British Interplanetary Society, 2010), p.85.
53 Pillinger, *My Life on Mars* (p.90).
54 Katharine Anderson, 'The Hydrographer's *Narrative*: Writing Global Knowledge in the 1830s', *Journal of Historical Geography,* 63 (2019), 48–60 (p.50).
55 Colin Pillinger, *Beagle – from Sailing Ship to Mars Spacecraft* (London: Faber and Faber, 2003).
56 National Maritime Museum Press Office, 'The Beagle Voyages – From Earth to Mars – A New Exhibition at the National Maritime Museum', *Royal Museums Greenwich Press Releases* (2002) https://www.rmg.co.uk/work-services/news-press/press-release/beagle-voyages-earth-mars-new-exhibition-national-maritime [11th August 2020].
57 Pillinger, *Beagle* (p.2).
58 Pillinger, *Beagle* (p.21, p.64).
59 Pillinger, *Beagle* (p.viii).
60 Jon Stratton, *Britpop and the English Music Tradition* (London: Routledge, 2010); Julian Stallabrass, *High art lite: British art in the 1990s* (London: Verso, 2001).
61 Edensor, *National Identity*.
62 Linda Colley, *Acts of Union and Disunion* (London: Profile, 2014).
63 Andy Bennett, '"Village Greens and Terraced Streets": Britpop and Representations of "Britishness"', *Young,* 5 (1997), 20–33 (p.25).
64 Tristan Weddigen, 'Alien Spotting: Damien Hirst's Beagle 2 Mars Lander Calibration Target and the Exploitation of Outer Space', in *Imagining Outer Space – European Astroculture in the Twentieth Century* ed. by Alexander Geppert (Basingstoke: Palgrave Macmillan, 2012), 304–315 (p.305).
65 Pillinger, *My Life on Mars* (p.135).
66 Weddigen, 'Alien Spotting' (p.308, p.305).
67 Weddigen, 'Alien Spotting' (p.306).
68 Pillinger, *My Life on Mars* (p.2).
69 René Bonnefoy and others, *Beagle 2 Commission of Inquiry* (ESA/UK, 2004), p.15.
70 Pillinger, *My Life on Mars* (p.219).

71 Billig, *Banal Nationalism* (p.4).
72 UK Space Agency, *Civil Space Strategy 2012–2016* (London: HMSO, 2012) p.2.
73 Lord Drayson, 'UK Space Agency Launch', *Department for Business, Innovation & Skills News and Speeches* (2010) https://web.archive.org/web/20100328163605/http://www.bis.gov.uk/News/Speeches/uk-space-agency-launch [30th May 2019].
74 Denis Cosgrove, 'Contested Global Visions: One-World, Whole-Earth, and the Apollo Space Photographs', *Annals of the Association of American Geographers,* 84 (1994), 270–294.
75 Lord Drayson, 'UK Space Agency Launch'. In 2009, Peake was recruited as one of six new ESA astronauts, adding to its existing pool that was recruited in the 1990s. There are currently fifteen serving members of the ESA Astronaut Corps.
76 Eric Seedhouse, *Tim Peake and Britain's Road to Space* (Chichester: Springer Praxis, 2017), p.53.
77 BBC News, 'Tim Peake launch: The seven Britons to go to space', *BBC News UK* (2015) https://www.bbc.co.uk/news/uk-35103788 [5th August 2020].
78 Department for Business, Innovation & Skills 'Tim Peake to be first British astronaut in space for more than 20 years', *Gov.uk, Science and Innovation* (2013): https://www.gov.uk/government/news/tim-peake-to-be-first-british-astronaut-in-space-for-more-than-20-years [31st May 2019].
79 Tim Peake, *Ask an Astronaut* (London: Penguin Random House, 2017), p.79.
80 Julien, 'Tim Peake's Spacewalk Overview' *European Space Agency* (2016): http://blogs.esa.int/tim-peake/2016/01/07/tim-peakes-spacewalk-overview/ [5th June 2019].
81 Jonathan Webb, Helen Briggs and Bernadette McCague, 'As it happened: Tim Peake's first spacewalk', *BBC News, Science and Environment* (2016): https://www.bbc.co.uk/news/live/science-environment-35303186/page/3 [5th June 2019].
82 Andrew Maclaren, 'Geopolitical Imaginaries of the Space Shuttle Mission Patches', *Geopolitics* (2019) [online].
83 Peake, *Ask an Astronaut* (p.149).
84 Derek P McCormack, 'Geographies for Moving Bodies: Thinking, Dancing, Spaces', *Geography Compass,* 2 (2008) 1822–1836.
85 BBC News, 'London Marathon 2016: Major Tim Peake launches race', *BBC News, London* (2016): https://www.bbc.co.uk/news/uk-england-london-36112994 [10th June 2019].
86 Seedhouse, *Tim Peake* (p.179).
87 Peake, *Ask an Astronaut* (p.135).
88 Peake, *Ask an Astronaut* (p.135).
89 Closs Stephens, 'The affective atmospheres' (p.181).
90 Closs Stephens, 'The affective atmospheres' (p.181).
91 European Space Agency, 'First Interview with Tim Peake back on Earth', *YouTube* (2016): https://youtu.be/Hgs16v_R38I [11th June 2019].
92 Science Museum Group, 'Tim Peake's Spacecraft Tour', *Science Museum Group* (2019): https://group.sciencemuseum.org.uk/our-work/tim-peakes-spacecraft-tour/ [11th June 2019].
93 Science Museum Group, 'Sokol KV-2 Emergency Suit', *Science Museum Group Collection* (2019): https://collection.sciencemuseum.org.uk/objects/co8614224/sokol-kv-2-emergency-suit-space-suit [12th June 2019].
94 Jaroslav Kalfar, 'A Memoir of a Year on the International Space Station', *New York Times* (2017) https://nyti.ms/2AZhPde [5th August 2020].

8 Conclusion

Diverse cultures, possible futures

This book started with a discussion of what it might mean to configure a geography of outer space in Britain from the start of the twentieth century to the present day. Prompted by the reflections of Arthur C Clarke and Denis Cosgrove on the ways in which science and the imagination might be considered together in theorising outer space, the book set out to consider perspectives on the extended cultural and political roots of spaceflight, the role of Britain in outer space, science and technology, and the possibilities of geographical enquiry into outer space that might echo older traditions in the discipline. The preceding chapters have demonstrated that there have been vibrant engagements with both the concept and the reality of space exploration in Britain during this period, in examples that have ranged from science-fictional stories to plans for interplanetary exploration, popular cultures of astronomy, mysterious sightings of UFOs, international collaboration in space research, prospects for interstellar flight, and media representations of actual space missions. This chapter draws together the significance of this analysis while presenting two final case studies that look towards the future of outer space in Britain.

A wide range of geographical encounters with outer space has been explored throughout the preceding chapters. Cutting across cultural and literary geographies, Chapter 2 considered the ways in which the emergent literature of science-fiction configured outer space in relation to a range of actual and imagined spaces in the early twentieth century, employing the narrative techniques of cognitive estrangement to configure new geographies of outer space that changed the ways in which the Universe has been understood ever since. Chapter 3 looked to the emergence of the British Interplanetary Society as a synthesis of technology and culture that was dependent on a series of alternative spaces of scientific activity, identifying the importance of localised and international geographies in the birth of astronautics in Britain. In examining popular cultures of outer space in post-war Britain, Chapter 4 identified a series of spatialities that together defined a broad range of outer-space imaginaries, from the situated geographies of night-sky and atmospheric observation to the projected futures of human space exploration, both utopian and dystopian, all inflected by the broader context of British culture and politics after the Second World War. Chapter 5, from the perspective of critical astropolitics, considered the geo-political cultures of outer space that shaped the British space programme in the post-war period, finding it to be entirely contingent on a shifting sense of Britain's place in the world. Chapter 6 explored the ways in which notions of interstellar travel have been connected to understandings of place and environment, applying such concepts to the possible other worlds of the galaxy, and the nature of the Solar System, while considering some of the implications of life across the galaxy, both human and alien. Focussing on three programmes of British space exploration, Chapter 7 found that concepts of nationalism continued to hold a strong grip on the cultures and politics of outer space in Britain, by means of national flagging, cultures of national memory, and performative nationalism. These chapters have traced a rich and varied narrative of cultural and political engagements with outer space that have foregrounded the importance of geography throughout.

The principal change that has been registered across the timeframe of this book has been the onset of spaceflight itself, while its anticipation and implications have been a consistent source of interest. This has not, however, been a simple story of linear progress, moving from imagination to reality in outer space. Indeed, scientific and technical knowledges have been seen to inform early imaginative representations of outer space in science fiction and other anticipatory texts, while imaginative geographies clearly had a role to play in later space missions and their reception. Central to this understanding have been the ways in which terrestrial places, spaces and landscapes have been significant in formulating understandings of outer space, and how humanity's engagement with outer space has, in turn, shaped a sense of place here on Earth, imaginatively and physically. Characteristic of these reciprocities has been the interplay between explorative and contemplative modes of engagement with outer space. As it has come to be appreciated that British spaceflight programmes, although significant in the mid-twentieth century and later, did not capture the broad imagination of the public in the ways that the American and Soviet programmes of that era did, so the significance of contemplative modes of engagement with outer space has become all the more apparent. In this way, a different set of stories about outer space has been told, less focussed on the immediate implications of spaceflight programmes, but enriched instead with imaginative cultural and political discourses of outer space, often inflected by the sublime as a mode of experience and representation. In this way, while British cultures and politics have informed understandings of outer space, understandings of outer space have also contributed to the making of British society itself since the start of the twentieth century.

In this concluding chapter, while reflecting on the findings of the book as a whole, two final case studies are presented that look towards the future of outer space in Britain and explain how a broad range of rich cultural and political discourses will continue to stir new ideas about the significance of outer space in Britain, including new exploratory visions of spaceflight launched from UK spaceports, as well as contemplative reflections in the form of the cosmic sublime in conceptual art. These examples crystallise many of the principal themes of the book, including the role of imaginative discourse in projections of spaceflight, the interplay between science and art in configurations of outer space, but above all the significance of geography in understanding and interpreting the cultures and politics of outer space in Britain.

UK spaceports: contested orbital landscapes

The past decade has seen a new global interest in establishing commercial spaceports for launching a diverse range of spacecraft. This signifies a shift away from national ownership of space-launch facilities, as a host of smaller, privately-backed facilities are emerging around the world, including Spaceport America outside Truth or Consequences, New Mexico; Equatorial Launch Australia in Arnhem, in Australia's Northern Territory; and Rocket Lab's Launch Complex 1 in Mahia, New Zealand. Such sites are currently in varying states of readiness to offer commercial satellite launches in the near future, with sub-orbital space tourism a more distant prospect. The UK is part of this global shift, with several small spaceports currently in development, representing an interest that has arisen partly out of the growth of UK satellite-manufacturing and space-services industries, alongside a more general proliferation and specialisation of satellite technologies since the late twentieth century.

The anticipated expansion of the commercial market for space launches in the UK should come as no surprise, as British companies, laboratories, and universities have long been involved in the space science and aerospace industries (see Chapters 5 and 7). While the commercialisation of outer space in the past few decades has been driven by the growth of satellite industries, particularly in the communications, environmental monitoring and surveillance sectors, these industries have relied on support from governments and the public sector. This new era in space exploration, with private companies operating alongside national and international space agencies, is occurring in both the Earth-orbital sphere, as well as increasingly in resource-oriented approaches to deep space exploration.[1] Moreover, the global satellite industry has been recognised as being closely associated with the neoliberal economic policies of Western states since the late twentieth century and related geopolitical imperatives.[2] One outcome of this association between industry and the state has been the application of satellite imagery to military as well as commercial uses, with these sectors said to be 'going hand in hand to assert national objectives in outer space'.[3] Indeed, despite landmark international agreements, such as the Outer Space Treaty (1967) and the Anti-Ballistic Missile Treaty (1972–2002), the orbital realm is increasingly seen as militarised and even weaponised.[4] An emerging regulatory and legislative environment has, therefore, created a new frontier for the commercialisation of outer space in Britain, with important implications for the militarisation of space.

In domestic political discourse, the establishment of the UK Space Agency in 2010 and the endorsement of Tim Peake as Britain's first 'official' astronaut in 2015 signalled a renewed focus on space exploration in Britain. As well as indicating a closer involvement with the European Space Agency's human spaceflight programme, this move has sought to carve out a greater role for the private space industry, including satellites but also in space science and prospective space tourism. The Space Industry Act in 2018 became the first major outer-space legislation in the UK since 1986, and identified satellite-launching as a growth industry of the future, estimated to be worth £10 billion to the UK economy.[5] Advocacy groups including the BIS have supported such moves enthusiastically, seeing the 2018 Act as a cornerstone of Britain's future as a spacefaring nation, while upholding the Society's long-standing ambitions for spaceflight (see Chapter 3). Such discourses are emblematic of a new phase of space exploration, that looks back on prior achievements in British spaceflight with a sense of nostalgia, while excitedly anticipating 'an indigenous launch capability for the UK'.[6] As such, recent *Spaceflight* editorials recall the British-made Prospero satellite of 1971 as an icon of past British achievement in space (see Chapter 5), while describing the UK as forcefully 'stand[ing] alone' and independently pursuing space exploration through partnerships between the public and private sectors, with the prospect of regular space launches taking place 'from British soil'.[7] Here, familiar national touchstones have again come to inform the ambitions of space exploration.

The prospect of establishing a home-grown satellite-navigation system to replace the UK's involvement in the European Union's Galileo project post-Brexit has acted as a further stimulus to the UK space industry. In July 2020, the UK Government bought a £400 million stake in the bankrupt American satellite firm OneWeb, a move designed to create a 'sovereign space capability' with applications across the commercial, communications and defence sectors.[8] Indeed, while Galileo will still be available to UK users for everyday satellite navigation, its more precise military and strategic positioning system is restricted to EU member-states. With the anticipation of a UK space strategy, amid such investments in the space industry, it is fair to say that the UK government is serious about space in the new commercial era. While there is a clear potential for economic growth in this area, concerns remain about increased military use of Earth-orbital space, as well as the wider political environment of nationalism and international competition in space. British spaceport proposals have further contributed to such narratives, enrolling distinct geographies of landscape, aerial and orbital space.

The Space Industry Act paved the way for the development of commercial spaceports in the UK for the first time. One immediate consequence of the Act was the establishment of the cross-departmental policy programme LaunchUK, through which private companies could bid for government funding towards the specific goal of creating the UK's first spaceport. Similar legislation in the United States has led to the Federal Aviation Administration approving eleven permits for commercial spaceports across the US by 2018.[9] The LaunchUK competition attracted a range of bids, including for a vertical-launch spaceport in Sutherland, on Scotland's north coast, as well as several options for horizontal-launch spaceports, including in Cornwall, Snowdonia and South Uist in the Outer Hebrides. While the vertical-launch option consists of staged rockets using conventional ballistic trajectories, the horizontal launch involves a modified jumbo jet taking off from a runway, with rocket stages deployed mid-flight to take payloads up into orbit. Each system has its own advantages, with the more ambitious vertical method offering the potential for greater payloads, including the possibility of multiple satellites per launch, and the horizontal method lower in cost, being able to rely on existing airfield infrastructure. Both approaches take advantage of coastal launch sites, with hardware from launch fallout, including booster stages and other abortive materials, able to crash-land in the seas around Britain. The Sutherland proposal was favoured in the first round of bidding, while the funding streams initiated by the Space Industry Act have encouraged continued development of spaceport plans across a range of options, with the Cornwall site, in partnership with Virgin Orbit, additionally awarded funding.

The emerging UK launch industry configures space in a variety of ways, from local landscapes to geopolitical positioning. This has involved envisaging the technological landscapes of the future, reformulating the value of peripheral spaces, and aligning with a variety of global orbital pathways. The qualities of this spatiality are typically summarised as the 'right geography' for the space-launch industry in the new commercial space age.[10] In a certain sense this space-launch geography updates prior anticipatory visions of global spaceports aligned with the territories of the former British Empire, from the mountain-tops of Africa to the high latitudes of Antarctica, in its projection of British national interests to outer space (see Chapter 5). In 2020, the proposers of the Sutherland site, the Scotland-based technology start-up Orbex and the defence multinational Lockheed Martin, were awarded £17.3 million in funding, sourced from Scotland's Highlands and Islands Enterprise, the UK's Nuclear Decommissioning Authority, and the UK Space Agency, and theirs is, at the time of writing, the most advanced of the proposals submitted to the LaunchUK scheme.[11] Prior to this, a government press release stated that 'Scotland is the best place in the UK to reach in-demand satellite orbits with vertically launched rockets', namely the polar orbit.[12] Across the worldwide launch industry, different kinds of satellite orbit serve different purposes, with far-out geostationary orbits mostly used for Earth-monitoring applications and global communications, as opposed to the low-Earth orbits which are more typically used for reconnaissance, mapping and navigation.[13] The Sutherland proposal involves launching satellites due north into a polar low-Earth orbit, which is beneficial in terms of its potential to cover the entire Earth's surface in as little as fourteen days, while crossing directly over both polar regions.[14] This approach takes advantage of Scotland's high latitudes, in contrast to tropical and equatorial sites which, due to the effects of the Earth's rotation, are more suited to attaining geostationary and other extended orbits. Accordingly, a LaunchUK promotional poster (Fig. 8.1) uses landscape imagery to implicitly favour sites in the far north such as Sutherland. It does this by portraying a vertical rocket launch against a mountainous backdrop, with the blue-green hues of the Aurora Borealis signifying northerly skies. In such ways, the spaces of outer space again become intimately connected to the terrestrial spaces of the Earth's surface.

The ability to efficiently attain polar orbit is something of a unique selling point of the Sutherland site. This was confirmed in an Orbex press-release video in August 2019, which claimed that 'Sutherland is one of the few locations in Europe to offer easy access to polar orbits'.[15] This raises questions about the potential for further militarisation of space, when considered in the broader context of the interplay between government, private finance and the defence industry, alongside the emerging geopolitics of the Arctic. The involvement of the American aerospace-defence contractor Lockheed Martin as part of the bidding consortium infers the likelihood of the Sutherland launch site being considered as 'dual-use', a designation that has been used to describe combined civilian and military applications of space hardware.[16] This adds to concerns about outer space being treated

Figure 8.1 LaunchUK promotional image

Source Credit: UK Space Agency

as 'the next frontier for military-neoliberal hegemony', in the renewed context of a broader range of nation-states becoming involved in space-launch activities.[17]

The UK Government's publication of a Defence Arctic Strategy in 2018 raises further questions about the militarisation of the Arctic region, including its volumetric spaces. This strategy, the first of its kind, has been understood as part of official responses to 'environmental, geopolitical and geo-economic trends affecting the Arctic', including the resurgence of Russia as a military power, the opening up of new shipping routes through the Arctic Ocean as a result of sea-ice melting, and the anticipation of new frontiers of resource extraction similarly exposed as a consequence of anthropogenic global warming.[18] It involves the active deployment of military hardware to Arctic spaces, including cold-weather training of ground troops, RAF patrols of Arctic airspace and the deployment of submarines under the Arctic sea ice.[19] This contributes to a sense in which the Arctic is understood a region that is gradually 'melting into the realm of "normal" politics', not only at sea-level but also above and below the surface, and illustrates the ways in which outer-space activities in the militarised realm often mirror grounded geopolitical realities.[20] While ostensibly acting as a forerunner of technological and economic progress in the UK space sector, the proposed Sutherland site, whether or not it ultimately becomes Britain's first spaceport, must be seen in this broader political context, as well as in the historical context of British cultures and politics of space exploration.

As well as having implications for nationalism in space and the further militarisation of space, spaceports have also attracted critical attention in other ways, including being understood through concepts of empire and utopia, tropes that have become well-known in critical studies of outer space.[21] The spaceport of the European Space Agency in Kourou, French Guiana, for example, has been interpreted as a neo-colonial installation, whose construction in the 1970s served to entrench colonialist inequalities, while Spaceport America, completed in 2011, has been questioned for its use of millions of dollars of US taxpayers' money, as its corporate tenants, including Virgin Galactic, have so far failed to launch their utopian-inflected space tourism projects.[22] Understandings of empire and utopia have also implicated notions of peripherality in the geographies of outer space. For example, the 'empty' Central Aboriginal Reserve of the Australian continent was materially overlooked by those planning the adjacent Woomera rocket range (see Chapter 5), while the penal colony that was established in French Guiana during the colonial era served as an indication of the value of this location to the imperial metropole. In such ways, the 'right geographies' of spaceport location often implicate places that are either seen as immaterial to national economies and political objectives, or available for exploitation.

Notions of peripherality have also been understood in environmental and social terms, and the anticipated spaceport for Sutherland in Scotland, which was granted planning permission in August 2020, has raised some controversy in terms of its likely impact on local landscapes. The A'Mhoine peninsula, where the spaceport is proposed, is part of the Caithness and Sutherland Peatlands Special Area of Conservation, and contains a Site of Special Scientific Interest, its blanket bog an important sink of carbon, and home to a breeding bird assemblage of national importance.[23] While, from the planners' perspective, the site is described as part of an 'untouched landscape' ripe for development, this landscape is valued in other ways as part of 'the far north flow country', whose ecological significance is noted and valued in social and cultural terms, as 'the birthright of Scots and the world'.[24] A petition opposing the spaceport proposal has attracted over one thousand signatories, with commentary tending to cite the ecological significance of the peninsula, the perceived lack of benefit to the local community, concerns about light and noise pollution, and the military ambitions associated with the site.[25]

Significantly, representations of landscape and the sublime are used in both sides of this debate, portraying A'Mhoine in contrasting terms, as one of the UK's last wilderness areas under threat from the space industry, or conversely as a high-tech landscape of progress and opportunity. As such, a host of computer-generated landscape images typically portray dramatic rocket launches against a recognisable highlands background as visions of the technological sublime, while landscape photography recalling the traditions of the Romantic sublime is used to illustrate protests against the development. Aspects of both can be seen in the LaunchUK promotional landscape composite (Fig. 8.1), in an image that synthesises the contested nature of sublime landscapes. In such ways, notions of the sublime have been incorporated into debates about outer-space futures in Britain, a theme that has become central to the understanding of outer space in a further realm of representation and experience, that of contemporary conceptual art.

Conceptual art and outer space: Katie Paterson and the cosmic sublime

Throughout this book, it has been recognised that representing and experiencing outer space has not only taken place through configuring acts of exploration and spaceflight. Indeed, as British efforts to establish space programmes have been limited in comparison to other national and international programmes, a diversity of imaginative discourses has come to the fore in British cultures and politics of outer space, some of which has involved the anticipation of spaceflight and some of which has found expression through more passive contemplations of the cosmic environment. While the notion of the sublime has been a factor in imagining space exploration, as noted in the above example, the sublime has been central to these alternative understandings of outer space, sometimes in opposition to the idea that outer space is to be explored, exploited and colonised, and this trope has been identified in various parts of this book (see Chapters 2 and 4). Speaking to this theme, a final case study considers the sublime in the works of the Scottish conceptual artist Katie Paterson, whose works reflect on the nature of outer-space phenomena and the human relationship with the cosmic environment. In setting some of these works in context, while recalling some of the preceding case studies of this book, the sublime is confirmed as a consistent touchstone for understandings of outer space in Britain since the start of the twentieth century.

Originally theorised in the writings of eighteenth-century philosophers Edmund Burke and Immanuel Kant, the sublime has come to represent scalar concepts of immensity, infinity, and deep time, while also aligning with certain emotions and affective resonances in the human body and mind, such as feelings of awe, fear and incomprehension.[26] One important influence on the emergence of the sublime in Western culture was the scientific revolution, during which new understandings of geological eras, the evolution of species, and the immensity of the Universe challenged many existing perceptions of time, space, and life itself. The sublime has been expressed in many ways through creative practice, particularly in the tradition of Romanticism that took hold in European and American imaginations during the eighteenth and nineteenth centuries. As part of this, the visual arts have pictured the sublime in various ways, most notably in the painted landscapes of artists such as J M W Turner and John Constable in the nineteenth century, who considered the sublime in nature, from the ferocity of raging storms to the beauty of sunlit meadows. Later iterations of the sublime in art have embraced a variety of forms, from the abstract expressionism of Mark Rothko to Damien Hirst's shark in formaldehyde. It is through such artistic, poetic, and philosophical traditions of the sublime that a distinctive set of engagements with outer space can be identified, noting also the sublime's association with that which is held aloft, or transcendent. In such ways, the sublime has become a proxy for the emotional impact of spaceflight in the modern age, and such frameworks help to understand outer space as a space of contemplation, perhaps engendering feelings of fear or reverence, and above all, a renewed sense of cosmic perspective.

The sublime and outer space have been understood together in various ways, in some cases combining scientific and artistic perspectives. Indeed, pictorial renderings of outer-space phenomena formed an integral part of scientific practices of observation and recording, especially when combined with telescopic sightings. This has been seen in lunar astronomy from Johannes Hevelius in the seventeenth century, to James Nasmyth in the nineteenth century, who were both expert draughtsmen as well as scientific observers.[27] These associations have continued into the Space Age, with astronomical imagery from space telescopes being understood in the context of the Romantic sublime in art.[28] More broadly, the visual arts have been understood as significant in shaping the 'social imaginary of outer space', and the sublime has represented a substantial part of this significance.[29] Elements of this can be seen in contemporary artworks such as James Turrell's 'Skyspaces', which entice viewers to behold vertical celestial spaces from within architectural and landscape installations, and Olafur Eliasson's 'The Weather Project', that filled London's Tate Modern with an immense glowing replica of the Sun.[30] In the American context, older traditions of the sublime in landscape imagery, from the Hudson River School to the photography of Ansel Adams, have acted both as a prompt for imagining the future exploration of space, as well as a means of subverting such imaginaries. As such, the mid-century science-fictional landscape paintings of Chesley Bonestell have been understood as extensions of the frontier sublime tradition to the imagined spaces of planetary conquest, whereas the situated landscape photographs of Trevor Paglen intend to problematise the surveillance activities of military satellites and other 'geographies of the black world' in twenty-first-century America.[31] In such ways, the sublime has acted as a potent field of representation in understanding the complex geographies of outer space.

In Britain, Katie Paterson has emerged as one of the most original voices in contemporary conceptual art. Her work has been exhibited in over thirty solo exhibitions and over a hundred group exhibitions in many countries around the world. Often taking a simple idea as a starting-point, Paterson's art combines detailed logistical, scientific, and technical research with sublime contemplations on themes such as the scale of the Universe, the nature of deep time, and the elemental make-up of stars and planets, considering both cosmic subjects and natural processes on Earth. As conceptual art, or the art of ideas, Paterson's works find meaning in the cognitive space between artist and audience and, like much of modern art, has moved beyond just pictorial representation to embrace a wide variety of artistic formats, including installations in gallery spaces but also artworks that are embedded in landscapes and the extended spaces of the natural environment. Installations in particular have been investigated for 'the critical spatial sensibilities instigated through [their] configuration of bodies, spaces and objects', acknowledging the involvement of affect and embodiment in the experience of such works.[32] Although Paterson's installations draw influence from post-war conceptual art, as well as land art, thematically her work is connected strongly to the tradition of the Romantic sublime. This connection was affirmed in a 2019 exhibition of her works alongside a selection of paintings by J M W Turner at the Turner Contemporary gallery in Margate, Kent.[33] As such, her works have been seen both to exploit the sublime and subvert it with 'humour and absurdity', creating 'a vertiginous sense of human relativity', a notion whose origins can be found in the imagined descriptions of cosmic ascent in some Classical and Renaissance works.[34] While recognising these deeper traditions, considering Paterson's works of the past decade can help bring meaning to broader cultures and politics of outer space in Britain, and serve to demonstrate the ways in which the sublime experience of cosmic space has evolved into the twenty-first century. Three works in particular resonate with this notion of the 'cosmic sublime', that have assembled scientific data about astronomical events, catalogued photographic imagery of deep space, and re-constituted an object from space.

All the Dead Stars (2009) is a map of outer space that documents every recorded instance of a star reaching the end of its life and exploding in a supernova event (Fig. 8.2). Over 27,000 such events in human history have so far been chronicled, mostly in recent years using specialist astronomical instruments, but also historically, and some with the naked eye. The vast majority of these recorded supernovae have occurred outside of the Milky Way galaxy. Each is represented in the artwork by a white dot on a black background, in the form of a three-by-two-metre, laser-etched aluminium panel. In constructing the artwork, Paterson collaborated with Professor Ofer Lahav, Chair of Astronomy at University College London, along with dozens of amateur 'supernova hunters' from across the world, to collate the data sets and configure them spatially.[35] The map itself reveals patterns that provoke reflections on the nature of the Universe and human relationships to it. Traces of a graticule, along what could be lines of latitude and longitude, can be observed in the supernova recordings, suggesting a regularity of observation patterns on Earth, while broader sweeping arrangements of dots denote the density of supernova observations in particular parts of the night sky. Such patterns are revealed through absence – the stars the supernovae extinguished no longer exist – while the title of the work denotes death and the reciprocal possibility of life in the cosmos. The life and death of stars and their conceptualisation as living entities resonates with certain

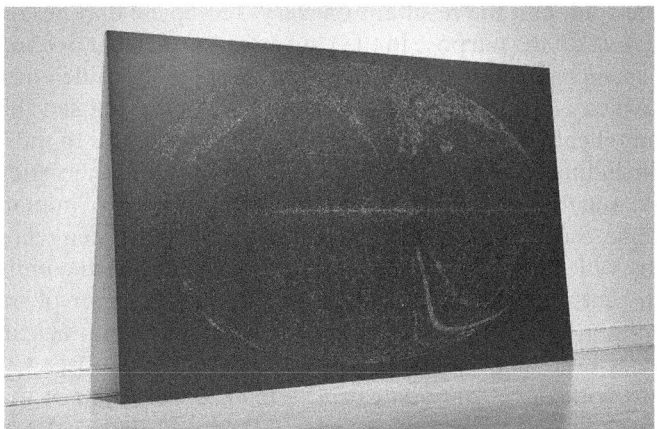

Figure 8.2 Katie Paterson, All the Dead Stars, 2009

Photo © Mead Gallery installation view, Mead Gallery, Warwick Arts Centre, 2013

Source Credit: Katie Paterson Studio

descriptions in Olaf Stapledon's *Star Maker*, which was written amid reg-ular observations of the night sky and a changing understanding of cos-mic evolution and deep time (see Chapter 2). The form of *All the Dead Stars* provokes comparisons with other mappings of interstellar space, such as the Daedalus research team's attempt to plot a course for future humanity through the stars (see Chapter 6). While sharing a novel use of astronomical data, Paterson's map differs fundamentally, in that it is not a course for future space travel, but a means for contemplation on the lim-ited life-span of cosmic entities, including perhaps humankind itself. In such ways Paterson's work can be seen to question the purpose of space exploration, promoting instead more melancholic reflections on the nature of the Universe.

History of Darkness (2010–) similarly deals with deep space and the consti-tution of the Universe. Inspired by the research of astronomers at Hawaii's Keck Observatory, it is an ongoing collection of thousands of images of cos-mic darkness, each annotated with a number representing its distance from Earth, from 100,000 to 13 billion light-years away, thereby assigning each slice of darkness a place measured in time as well as distance. The images them-selves, in the form of 35mm photographic slides, are also numbered on a scale from one to infinity, playing with the convention of multiple editions of a par-ticular artwork. The work acknowledges darkness as a principal characteristic of outer space, a quality that has perhaps been forgotten in the modern pur-suit of light and enlightenment.[36] Paterson's preoccupation with darkness is said to invoke 'the apocalyptic and the inevitable dissolution of the universe', while her repeated evocation of deep time, infinity and unimaginable scale has been noted to 'induce a frisson of terror' in the observer.[37] The system-atic nature of her cataloguing of darkness also suggests a search for meaning in the cosmic void, and the resultant imagery – complete darkness – perhaps denotes a lack of divine purpose in the Universe. As such, her documenting of interstellar space has diverted data from the quest for scientific knowledge to something which is 'felt but not understood, known but not sensible, shown but not comprehendible'.[38] Aspects of the sublime are central to the impact of this work, including a sense of awe and incomprehension at the extent of cos-mic darkness and the feeling of fear or terror at the prospect of human solitude in the Universe. Sublime understandings such as this have been recognised as the means by which certain cultures of outer space have been communicated in twentieth-century Britain, including the anticipated terror of malevolent alien life in literary imaginations (see Chapter 2) and the ontological paradox at the heart of UFO contact narratives in post-war Britain (see Chapter 4). Paterson's artwork thereby animates and extends conceptions of the sublime in outer-space cultures in Britain.

Campo del Cielo (2012–14) consists of a small metallic meteorite that has been melted, re-cast into its original shape, and sent back into space. The re-casting process fundamentally altered the chemical structure of the meteorite, although its elemental make-up and shape were identical to its original state. It was sent to the International Space Station on board an automated supply vehicle operated by the European Space Agency, which was then returned, along with waste material from the ISS, burning up in the Earth's atmosphere in a planned destructive re-entry procedure. This act of object performance is part of the process of 'making the strange familiar' that permeates much of Paterson's work.[39] In its enactment, the artwork 'reverses a terracentric perspective' in what has been described as a 'Copernican gesture', de-centring the Earth from understandings of the Universe.[40] This sense of re-alignment of humanity's place in the cosmos further picks up on themes identified in the literary geographies of Stapledon and Wells, whose works foregrounded the multiplicity of worlds and the cosmic downgrading of humanity in the context of a Universe filled with potential forms of other life (see Chapter 2). Furthermore, in returning a celestial object back from whence it came, and in the ultimate intended destruction of the object, *Campo del Cielo* subverts many of the dominant imperatives of spaceflight that have been established over the past century, including the British Interplanetary Society's utopian-inflected plans for space exploration, and performances of nationalism in space exploration (see Chapters 3, 5 and 7). This inversion of expectation questions the underlying motivations of space exploration, presenting it as a circular, almost redundant exercise that is futile with respect to the immensity of the cosmic realm.

In Paterson's work, represented by these three examples, a broad synthesis of alternative understandings of outer space can be identified, drawing from the sublime, while combining techniques from across the arts and science. In this way, the recording of astronomical data has provided a way of contemplating life and death in the Universe, while the appreciation of cosmic darkness has provoked reflections on the constitution of interstellar space, and the manipulation of a metallic meteorite has queried the existing conventions of spaceflight. Such understandings can help to bring together work in geography that has explored the embodied and affective meanings of installation and landscape art, with an appreciation of the sublime, and thereby deepen understandings of how people have connected with outer space through the experience of the sublime. As such, while understandings of art and landscape in geography have in the past favoured the visual as a route to understanding, a renewed appreciation of the ways in which people 'move in, feel for, and are affected by, the material world' helps to enrol a fuller range of emotive and sensual understandings of the geographies of outer space.[41] These examples also illustrate how the sublime has had the capacity to invert or displace the anticipation of space exploration, as defined by certain groups and individuals who promoted spaceflight in the twentieth century. As such, Paterson's work, and the sublime more broadly, will continue to inspire deep contemplations of cosmic spaces and reflective considerations of humanity's role in the Universe, drawing from the basic imperative to look up at the night sky that has long preoccupied the human mind.

Conclusion: the shape of things to come?

Throughout this book, a series of reciprocal relationships have been instrumental in understanding the geographies of outer space in Britain. These have been between science and the imagination, terrestrial space and cosmic space, and explorative and contemplative understandings of outer space. This book has helped to demonstrate that, rather than standing in opposition, as separate cultures of outer space, these interfaces, and the geographies that are implicated by them, have been where the most significant and enduring configurations of outer space culture have been found, and where the fundamental relationships between Earth, cosmos and culture have been illuminated. This reflects geography's capacity to bring together that which is usually kept apart, and in such ways can outer space be understood as a dynamic focus of enquiry in geography, considered as a cultural space, a political space, and as a discursive space of contention in society, as well as a space of exploration, science and discovery.

Across all of these themes and forms of understanding, outer space will continue to captivate people in Britain and around the world into the future. Several engagements are likely to proceed into the 2020s, which will probably see the UK's first commercial spaceport being established, perhaps in Sutherland or Cornwall. This could usher in a new era of British space exploration, with satellite launches devoted to commercial and possibly military applications becoming a regular occurrence from the British landmass. The advent of space tourism is a less distinct, but nonetheless possible scenario that may become available to a small, wealthy section of society. Away from the prospect of space launches, British industries, supported by the UK Space Agency, will continue to be involved in outer-space research and the servicing of the global space sector, in collaboration with a range of international agencies and contractors. It is also likely that Tim Peake, or another British astronaut, will extend the short but significant sequence of Britons to have visited orbital space stations. In terms of an evolving cultural understanding of outer space, there remain ample opportunities for narrative and other imaginative engagements with cosmic subjects, that will draw on the UK's rich cultural histories as well as forge new types of representation and experience. The prime-time 2019 television adaptation of *The War of the Worlds* by the BBC is a reminder of the enduring legacy of H G Wells, while Olaf Stapledon's *Last and First Men* is also the subject of a new cinematic interpretation in 2020 by the Icelandic composer Jóhann Jóhannsson, to give just two very recent examples of this continued re-imagining of older narratives. A vibrant museum culture will further foster the interpretation of historic and contemporary understandings of outer space, with major exhibitions at the National Maritime Museum and the Science Museum already providing the headline acts, and with outer-space exhibits continuing to act as mainstay permanent features of science museums, alongside a host of smaller engagements across the country, ensuring that cultures of outer space continue to reach wide audiences in the UK. Questions around the cultural and political diversity of exhibitions and audiences, as well as the necessity of decolonial approaches to representations of space exploration, will remain important in such engagements with outer space.

With outer space likely to remain in the British public consciousness in such ways, there is a continued need for critical interpretations of outer-space cultures and politics, and the geographical approaches explored in this book offer one set of techniques for possible scholarly engagement. In recent years, however, apart from a few exceptions, there has been a lack of critical studies in geography, compared to other allied disciplines, on outer space. Enabled by the breadth of the discipline, this book has been able to cover cultural, political, environmental and historical geographies of outer space in Britain, while considering a range of theoretical approaches to geographical knowledge through studies of representation, audiences, nationalism, affect, and volumetric understandings of space, underpinned by geography's enduring synthesis of scientific and societal understandings of the Earth and cosmos together. In such ways, this book has attempted to show that geography can be a natural home for critical studies of outer space. Yet there remain many ways in which the geographies of outer space could be extended. How, for example, might the socio-political inequalities in the UK space industry be explored from a geographical perspective? In what ways do the contemporary amateur astronomers emotionally engage with cosmic concepts of the sublime? And how have the narrative geographies of science-fictional texts been read and understood by audiences in different times and places? Understanding geographies of outer space in such ways provides unique insights on the relationships between humankind and the cosmos, and many such opportunities present themselves for further insight. It is hoped that further studies will contribute in such areas and help to revive the historic connection between Earth and cosmos in geographical enquiry.

Notes

1 Jason Beery, 'State, Capital and Spaceships: A Terrestrial Geography of Space Tourism', *Geoforum,* 43 (2012), 25–34; Julie Klinger, *Rare Earth Frontiers – from Terrestrial Subsoils to Lunar Landscapes* (Ithaca: Cornell University Press, 2017).

2 Jocelyn Wills, 'State Surveillance and Outer-Space Capitalism: The Case of Macdonald, Dettwiler and Associates', in *The Palgrave Handbook of Society, Culture and Outer Space*, ed. by Peter Dickens and James S Ormrod (Basingstoke: Palgrave Macmillan, 2016), 94–122.

3 Barney Warf, 'Geopolitics of the Satellite Industry', *Tijdschrift voor Economische en Sociale Geografie,* 98 (2007), 385–397; Patrick Salin, 'Privatization and Militarization in the Space Business Environment', *Space Policy,* 17 (2001), 19–26 (p.19).

4 Alexander Geppert, Tilmann Seiebeneichner and Daniel Brandau, eds, *Militarizing Outer Space: Astroculture, Dystopia and the Cold War* (London: Palgrave Macmillan, 2020); Joan Johnson-Freese, and David Burbach, 'The Outer Space Treaty and the Weaponization of Space', *Bulletin of the Atomic Scientists,* 75 (2019), 137–141.

5 Department for Transport, 'New laws unlock exciting space era for UK', *Gov.uk News* (2018) https://www.gov.uk/government/news/new-laws-unlock-exciting-space-era-for-uk [25th October 2019].

6 'Out to Launch', *Spaceflight*, 60, 9 (2018), p.4.

7 'A Ray of Hope?', *Spaceflight*, 60, 10 (2018), p.5; 'Britain in Space: Up, Up and Away', *Spaceflight*, 60, 5 (2018), p.7.

8 BBC, 'UK government takes £400m stake in satellite firm OneWeb', *BBC News – Science and Environment* (2020) https://www.bbc.co.uk/news/science-environment-53279783 [22nd August 2020].

9 Katherine Sammler and Casey Lynch, 'Spaceport America: Contested Offworld Access and the Everyman Astronaut', *Geopolitics* (2019).

10 Department for Transport, 'How we are promoting and regulating space-flight from the UK', *Gov.uk Science and Innovation* (2019) https://www.gov.uk/guidance/how-we-are-promoting-and-regulating-spaceflight-from-the-uk [28th October 2019].

11 BBC News, 'Sutherland Space Hub secures planning permission', *BBC News – Scotland* (2020) https://www.bbc.co.uk/news/uk-scotland-highlands-islands-53834962 [21st August 2020].

12 Department for Transport, 'One Giant Leap: Vertical launch spaceport to bring UK into new Space Age', *Gov.uk Spaceflight* (2018) https://www.gov.uk/government/news/one-giant-leap-vertical-launch-spaceport-to-bring-uk-into-new-space-age [24th October 2019].

13 Christy Collis, 'The Geostationary Orbit: A Critical Legal Geography of Space's Most Valuable Real Estate', in *Space Travel and Culture: From Apollo to Space Tourism*, ed. by Martin Parker and David Bell (Oxford: Blackwell, 2009), 26–47.

14 Wills, 'State Surveillance'.

15 Orbex, 'Video animation of satellite launch from Scotland', *Orbex News* (2019): https://orbex.space/news/video-animation-of-satellite-launch-from-scotland [4th November 2019].

16 Roger Handberg, 'Dual-Use as Unintended Policy Driver: The American Bubble', in *Societal Impact of Spaceflight*, ed. by Roger D Launius and Steven J Dick (Washington DC: NASA, 2008), 353–368.

17 Fraser MacDonald, 'Anti-Astropolitik – outer space and the orbit of geography', *Progress in Human Geography* 31 (2007), 592–615 (p.594).

18 Duncan Depledge, Klaus Dodds and Caroline Kennedy-Pipe, 'The UK's Defence Arctic Strategy – Negotiating the Slippery Geopolitics of the UK and the Arctic', *The RUSI Journal,* 164 (2019), 28–39 (p.28).

19 Ministry of Defence, 'UK Government 'Defence Secretary announces new Defence Arctic Strategy', *Gov.uk Defence and Armed Forces* (2018): https://www.gov.uk/government/news/defence-secretary-announces-new-defence-arctic-strategy [6th November 2019].

20 Johanne M Bruun and Ingrid A Medby, 'Theorising the Thaw: Geopolitics in a Changing Arctic', *Geography Compass* 8 (2014), 915-929 (p.916).

21 De Witt Douglas Kilgore, *Astrofuturism: Science, Race and Visions of Utopia in Space* (Philadelphia: University of Pennsylvania Press, 2003); Denis Cosgrove, *Apollo's Eye: A Cartographic Genealogy of the Earth in the Western Imagination* (Baltimore: John Hopkins University Press, 2001).

22 Sammler and Lynch, 'Spaceport America'; Peter Redfield, *Space in the Tropics – from Convicts to Rockets in French Guiana* (Berkeley: University of California Press, 2000).

23 Sitelink, 'A'Mhoine SSSI Overview', *Sitelink* (2019) https://sitelink.nature.scot/site/1 [31st October 2019].

24 BBC, 'Sutherland spaceport plans cover "extensive" site', *BBC News – Scotland* (2019) https://www.bbc.co.uk/news/uk-scotland-highlands-islands-49905163 [22nd August 2020]; Ali G, 'Protect A'Mhoine peninsula's unique habitat from development as a rocket launchpad', *Care2 Petitions* (2019) https://www.thepetitionsite.com/en-gb/663/777/866/protect-amhoine-peninsulas-unique-habitat-from-development-as-a-rocket-launchpad/ [20th August 2020].

25 Ali G, 'Protect A'Mhoine peninsula's unique habitat'.

26 Philip Shaw, *The Sublime* (London: Routledge, 2006).

27 Francis Robertson, 'Science and Fiction: James Nasmyth's Photographic Images of the Moon', *Victorian Studies,* 48 (2006), 595–623.

28 Elizabeth Kessler, *Picturing the Cosmos – Hubble Space Telescope Images and the Astronomical Sublime* (London: University of Minnesota Press, 2012).

29 Nicola Triscott, 'Transmissions from the Noosphere: Contemporary Art and Outer Space', in *The Palgrave Handbook of Society, Culture and Outer Space*, ed. by Peter Dickens and James S Ormrod (New York: Palgrave Macmillan, 2016), 414–444 (p.414).

30 Saskia Warren, 'Audiencing James Turell's Skyspace: Encounters between Art and Audience at Yorkshire Sculpture Park', *cultural geographies*, 20 (2012), 83–102; Isabelle Loring Wallace, 'Technology and the Landscape: Turner, Pfeiffer and Eliasson after the Deluge', *Visual Culture in Britain*, 12 (2011), 57–75.

31 Daniel Sage, *How Outer Space Made America: Geography, Organization and the Cosmic Sublime* (Farnham: Ashgate, 2014); Henrik Gustafsson, 'Foresight, Hindsight and State Secrecy in the American West: The Geopolitical Aesthetics of Trevor Paglen', *Journal of Visual Culture*, 12 (2013), 148–164 (p.159).

32 Harriet Hawkins, '"The Argument of the Eye?" The Cultural Geographies of Installation Art', *cultural geographies,* 17 (2010), 321–340 (p.323).

33 Nick Alfrey, 'A Place That Exists Only in Moonlight: Katie Paterson and J M W Turner', *Turner Society News* 123 (Autumn 2019).

34 Ben Tufnell, 'Katie Paterson, Sand and Stars', *Katie Paterson, Texts* (2014) http://katiepaterson.org/texts/ [15th November 2019].

35 Mary Jane Jacob, 'The *Gedankenexperiments* of Katie Paterson', *Katie Paterson, Texts* (2016): http://katiepaterson.org/texts/ [19th November 2019].

36 Tim Edensor, *From Light to Dark – Daylight, Illumination and Gloom* (Minneapolis: University of Minnesota Press, 2017).

37 Tufnell, 'Katie Paterson'.

38 Sally O'Reilly, 'Of Great Magnitudes and Multiplicities', *Katie Paterson, Texts* (2016): http://katiepaterson.org/texts/ [19th November 2019].

39 O'Reilly, 'Of Great Magnitudes'.

40 Lars Bang Larsen, 'Astronomy Domine: The Anthropological-Cosmological Squeeze in Katie Paterson's work', *Katie Paterson, Texts* (2016): http://katiepaterson.org/texts/ [19th November 2019].

41 Nina Morris, 'Night Walking: Darkness and Sensory Perception in a Night-Time Landscape Installation', *cultural geographies,* 18 (2011), 315–342 (p.318); Denis Cosgrove, *Geography and Vision* (London: I B Tauris, 2008).

Bibliography

Research monographs, biographies, edited collections, and journal articles

Adey, Peter, *Aerial Life: Spaces, Mobilities, Affects* (Oxford: Wiley, 2010).

————Shining Eyes Are Dreaming and Their Dreams Are Wings": Affect, Airmindedness and the Birth of the Aerial Subject'"Ten Thousand Lads with, *cultural geographies* 18 (2010), 63–89.

Agar, Jon, *Science and Spectacle – The Work of Jodrell Bank in Post-War British Culture* (Amsterdam: Harwood Academic Publishers, 1998).

Agnew, John, and Gearóid Ó Tuathail, 'Geopolitics and Discourse – Practical Geopolitical Reasoning in American Foreign Policy', *Geography*, 11 (1992), 190–204.

Alfrey, Nick, 'A Place That Exists Only in Moonlight: Katie Paterson and J M W Turner', *Turner Society News* 123 (Autumn 2019).

Anderson, Benedict, *Imagined Communities – Reflections on the Origins and Spread of Nationalism* (London: Verso, 1983).

Anderson, Katharine, 'The Hydrographer's Narrative: Writing Global Knowledge in the 1830s', *Journal of Historical Geography*, 63 (2019), 48–60.

Ashcroft, Bill, Gareth Griffiths, and Helen Tiffin, *The Empire Writes Back: Theory and Practice in Post-Colonial Literatures* (London: Routledge, 1989).

Ashley, Mike, *The Time Machines – The Story of the Science-Fiction Pulp Magazines from the Beginning to 1950* (Liverpool: Liverpool University Press, 2000).

Atkinson, David, Peter Jackson, David Sibley, and Neil Washbourne, eds. *Cultural Geography – A Critical Dictionary of Key Concepts* (London: I B Tauris, 2005).

Aubin, David, Charlotte Bigg, and H Otto Sibum, eds. *The Heavens on Earth – Observatories and Astronomy in Nineteenth-Century Science and Culture* (Durham, NC: Duke University Press, 2010).

Barnett, Clive, '"A Choice of Nightmares": Narration and Desire in Heart of Darkness', *Gender, Place and Culture*, 3 (1996), 277–292.

Barton, Ruth, *The X Club – Power and Authority in Victorian Science* (Chicago: University of Chicago Press, 2018).

Beery, Jason, 'State, Capital and Spaceships: A Terrestrial Geography of Space Tourism', *Geoforum*, 43 (2012), 25–34.

Bennett, Andy '"Village Greens and Terraced Streets": Britpop and Representations of "Britishness"', *Young*, 5(1997), 20–33.

Berger, Albert I, 'Science-Fiction Critiques of the American Space Program', *Science Fiction Studies*, 15 (1978).

Bleiler, Everett Franklin, *The Gernsback Years – A Complete Coverage of the Genre Magazines Amazing, Astounding, Wonder and Others from 1926 Through 1936* (Kent, Ohio: Kent State University Press, 1998).

Bloom, Ursula, *He Lit the Lamp – A Biography of Professor A M Low* (London: Burke, 1958).

Boes, Tobias, 'Beyond Whole Earth: Planetary Mediation and the Anthropocene', *Environmental Humanities*, 5 (2014), 155–170.

Boon, Timothy, *Films of Fact: A History of Science in Documentary Films and Television* (London: Wallflower, 2008).

Bormann, Natalie, and Michael Sheehan, eds. *Securing Outer Space* (New York: Routledge, 2009).

Bould, Mark, Andrew M Butler, and Sherryl Vint., eds. *Fifty Key Figures in Science Fiction* (New York: Routledge, 2009).

Bowler, Peter, *A History of the Future: Prophets of Progress from H G Wells to Isaac Asimov* (Cambridge: Cambridge University Press, 2017).

⸻'Experts and Publishers: Writing Popular Science in Early-Twentieth Century Britain, Writing Popular History of Science Now', *British Journal for the History of Science*, 39 (2006), 159–187.

⸻'Parallel Prophecies: Science Fiction and Futurology in the Twentieth Century', *Osiris*, 34 (2019), 121–138.

Breuer, William, *Race to the Moon – America's Duel With the Soviets* (Westport: Praeger, 1993).

Bruun, Johanne M, and Ingrid A Medby, 'Theorising the Thaw: Geopolitics in a Changing Arctic', *Geography Compass*, 8 (2014), 915–929.

Bulkley, Rip, 'Harbingers of Sputnik: The Amateur Radio Preparations in the Soviet Union', *History and Technology*, 16 (1999), 67–102.

Calder, Angus, *The Myth of the Blitz* (London: Pimlico, 1991).

Canavan, Gerry, '"A Dread Mystery, Compelling Adoration": Olaf Stapledon, Star Maker, and Totality', *Science Fiction Studies*, 43 (2016), 310–330.

Chapman, James, 'Onward Christian Spacemen: Dan Dare – Pilot of the Future as British Cultural History', *Visual Culture in Britain*, 9 (2008), 55–79.

Chatwin, Andrew, ed. *Val Cleaver (1917–1977) – A Very English Rocketeer* (London: British Interplanetary Society, 2014).

Christianson, Gale E, 'Kepler's Somnium: Science Fiction and the Renaissance Scientist', *Science Fiction Studies*, 3 (1976), 79–90.

Clarke, Arthur C, *Astounding Days: A Science Fictional Autobiography* (London: Victor Gollancz Ltd., 1989).

Clarke, David, *The UFO Files* (London: Bloomsbury, 2012).

Closs Stephens, Angharad, 'The Affective Atmospheres of Nationalism', *cultural geographies*, 23 (2016), 181–98.

Colley, Linda, *Acts of Union and Disunion* (London: Profile, 2014).

Collins, Martin, *Space Race – The US-USSR Competition to Reach the Moon* (Washington DC: Smithsonian, 1999).

Collis, Christy, and Klaus Dodds, 'Assault on the Unknown: The Historical and Political Geographies of the International Geophysical Year (1957–1958)', *Journal of Historical Geography*, 34 (2008), 555–573.

Conekin, Becky, *'The Autobiography of a Nation' – the 1951 Festival of Britain* Manchester: Manchester University Press, 2003.

Constemis, Athena, and Thérèse Encrenaz, *Life Beyond Earth – The Search for Habitable Worlds in the Universe* (Cambridge: Cambridge University Press, 2013).

Cook, John R, and Peter Wright, eds. *British Science Fiction Television – A Hitchhiker's Guide* (London: I B Tauris, 2006).

Correll, Randall R, and Simon P Worden, 'The Demise of US Spacepower: Not With a Bang but a Whimper', *Astropolitics: The International Journal of Space Politics & Policy*, 3 (2005), 233–264.

Cosgrove, Denis, *Apollo's Eye: A Cartographic Genealogy of the Earth in the Western Imagination* (Baltimore: John Hopkins University Press, 2001).

————'Contested Global Visions: One-World, Whole-Earth, and the Apollo Space Photographs', *Annals of the Association of American Geographers*, 84(1994), 270–294.

————*Geography and Vision*(London: I B Tauris, 2008)

Cosgrove, Denis, and Veronica Della Dora, eds. *High Places – Cultural Geographies of Mountains, Ice and Science* (London: I B Tauris, 2008).

Craggs, Ruth, and Martin Mahony, 'The Geographies of the Conference', *Geography Compass*, 8 (2014), 414–430.

Crossley, Robert, ed. *An Olaf Stapledon Reader* (New York: Syracuse University Press, 1997).

————'Famous Mythical Beasts: Olaf Stapledon and H G Wells', *Georgia Review* 36 (1982) 619–635.

————*Imagining Mars: A Literary History* (Middletown CT: Wesleyan University Press, 2011)

————*Olaf Stapledon – Speaking for the Future* (Liverpool: Liverpool University Press, 1994).

Danahay, Martin, 'Wells, Galton and Biopower: Breeding Human Animals', *Journal of Victorian Culture*, 17 (2012), 468–479.

Daniels, Stephen, and Catherine Nash, 'Lifepaths: Geography and Biography', *Journal of Historical Geography*, 30 (2004), 449–458.

Degroot, Dagomar, '"A Catastrophe Happening in Front of Our Very Eyes": The Environmental History of a Comet Crash on Jupiter', *Environmental History*, 22 (2017), 23–49.

Depledge, Duncan, Klaus Dodds, and Caroline Kennedy-Pipe, 'The UK's Defence Arctic Strategy – Negotiating the Slippery Geopolitics of the UK and the Arctic', *The RUSI Journal*, 164 (2019), 28–39.

Desmond, Adrian, 'Redefining the X Axis: "Professionals", "Amateurs" and the Making of Mid-Victorian Biology – A Progress Report', *Journal of the History of Biology*, 34 (2001), 3–50.

Dewan, William J, '"A Saucerful of Secrets": An Interdisciplinary Analysis of UFO Experiences', *The Journal of American Folklore*, 119 (2006), 184–202.

Dick, Steven J, and Roger D Launius, eds. *Societal Impact of Spaceflight* (Washington DC: NASA, 2007).

Dickens, Peter, and James S Ormrod, eds. *The Palgrave Handbook of Society, Culture and Outer Space* (Basingstoke: Palgrave Macmillan, 2016).

Dittmer, Jason, 'Colonialism and Place Creation in "Mars Pathfinder" Media Coverage', *The Geographical Review*, 97 (2007), 112–30.

Dittmer, Jason, ed. *Comic Book Geographies* (Stuttgart: Franz Steiner, 2014).

Dolman, Everett C, *Astropolitik: Classical Geopolitics in the Space Age* (London: Frank Cass, 2002).

Domosh, Mona, Michael Heffernan, and Charles W J Withers, eds. *The SAGE Handbook of Historical Geography* (London: SAGE, 2020).

Dougherty, Kerrie, 'Spaceport Woomera: The Anglo-Australian Vision of Woomera Rocket Range', *Quest – The History of Spaceflight*, 22 (2008), 1–19.

Duncan, James, Nuala Johnson, and Richard Schein. *A Companion to Cultural Geography* (Oxford: Blackwell, 2008).

Dunnett, Oliver, 'Geopolitical Cultures of Outer Space: The British Interplanetary Society, 1933–1965', *Geopolitics*, 22 (2017), 452–473.

Dunnett, Oliver, 'Imperialism, Technology and Tropicality in Arthur C Clarke's Geopolitics of Outer Space', *Geopolitics* (2019).

Dunnett, Oliver, 'Patrick Moore, Arthur C Clarke and "British Outer Space" in the Mid-Twentieth Century', *cultural geographies*, 19 (2012), 505–522.

Dunnett, Oliver, and others, 'Geographies of Outer Space: Progress and New Opportunities', *Progress in Human Geography*, 43 (2019), 314–336.

Edensor, Tim, *From Light to Dark – Daylight, Illumination and Gloom* (Minneapolis: University of Minnesota Press, 2017).

_____*National Identity, Popular Culture and Everyday Life* (Oxford: Berg, 2002)

Edgerton, David, *England and the Aeroplane – An Essay on a Militant and Technological Nation* (London: Macmillan, 1991).

_____*Warfare State: Britain, 1920–1970* (Cambridge: Cambridge University Press, 2006).

Eghigian, Greg, 'Making UFOs Make Sense: Ufology, Science, and the History of Their Mutual Mistrust', *Public Understanding of Science*, 26 (2017), 612–626.

Elden, Stuart, 'Secure the Volume: Vertical Geopolitics and the Depth of Power', *Political Geography*, 34 (2013), 35–51.

Farry, James, and David A Kirby, 'The Universe Will Be Televised: Space, Science, Satellites and British Television Production, 1946–1969', *History and Technology*, 28 (2012), 311–333.

Finnegan, Diarmid, 'Finding a Scientific Voice: Performing Science, Space and Speech in the 19th Century', *Transactions of the Institute of British Geographers*, 42 (2017), 192–205.

Forgan, Sophie, and Graeme Gooday, 'Constructing South Kensington: The Buildings and Politics of T H Huxley's Working Environments', *The British Journal for the History of Science*, 29 (1996), 435–468.

Foucault, Michel, 'Of Other Spaces', *Diacritics*, 16 (1986), 22–27.

France, Martin E B, 'Back to the Future: Space Power Theory and A T Mahan', *Space Policy*, 16 (2000), 237–241.

Garvía, Roberto, *Esperanto and Its Rivals – The Struggle for an International Language* (Philadelphia: University of Pennsylvania Press, 2015).

Geppert, Alexander, ed. *Imagining Outer Space – European Astroculture in the Twentieth Century* (New York: Palgrave Macmillan, 2012).

_____ed. *Limiting Outer Space: Astroculture After Apollo* (London: Palgrave Macmillan).

_____'Extraterrestrial Encounters: UFOs, Science and the Quest for Transcendence, 1947–1972', *History and Technology*, 28 (2012), 335–362.

_____'Space Personae: Cosmopolitan Networks of Peripheral Knowledge, 1927–1957', *Journal of Modern European History*, 6 (2008), 262–286.

Geppert, Alexander, Tilmann Seiebeneichner, and Daniel Brandau, eds. *Militarizing Outer Space: Astroculture, Dystopia and the Cold War* (London: Palgrave Macmillan, 2020).

Gibson, Chris, 'Cartographies of the Colonial/Capitalist State: A Geopolitics of Indigenous Self-Determination in Australia', *Antipode*, 31 (1999), 45–79.

Glendening, John, *The Evolutionary Imagination in Late-Victorian Novels: An Entangled Bank* (Aldershot: Ashgate, 2007).

Godwin, Matthew, *The Skylark Rocket – British Space Science and the European Space Research Organisation 1957–1972* (Paris: Beauchesne, 2007).

Gorman, Alice, 'La terre et l'espace: Rockets, Prisons, Protests and Heritage in Australia and French Guiana', *Archaeologies: Journal of the World Archaeological Congress*, 3 (2007), 153–168.

Green, Roger Lancelyn, *Into Other Worlds – Space-Flight in Fiction, from Lucian to Lewis* (London: Abelard-Schuman, 1958).

Gregory, Alan P R, *Science Fiction Theology: Beauty and the Transformation of the Sublime* (Waco, Texas: Baylor University Press).

Gustafsson, Henrik, 'Foresight, Hindsight and State Secrecy in the American West: The Geopolitical Aesthetics of Trevor Paglen', *Journal of Visual Culture*, 12 (2013), 148–164.

Hawkins, Harriet, '"The Argument of the Eye?' The Cultural Geographies of Installation Art", *cultural geographies*, 17 (2010), 321–340.

Higgitt, Rebekah, *Recreating Newton: Newtonian Biography and the Making of Nineteenth-Century History of Science* (London: Pickering and Chatto, 2007).

Hodder, Jake, Stephen Legg, and Michael Heffernan, 'Introduction: Historical Geographies of Internationalism, 1900–1950', *Political Geography*, 49 (2015), 1–6.

Holzmann Kevles, Bettyann, *Almost Heaven – The Story of Women in Space* (Cambridge, MA: MIT Press, 2006).

Hones, Sheila, 'Literary Geographies, Past and Future', *Literary Geographies*, 1 (2015), 1–5.

———'Literary Geography: Setting and Narrative Space", *Social and Cultural Geography*, 12 (2011), 685–699.

Hunter, Ian Q, ed. *British Science Fiction Cinema* (London: Routledge, 1999).

Jacobs, Jason, *The Intimate Screen: Early British Television Drama* (Oxford: Oxford University Press, 2000).

Jasanoff, Sheila, and Sang-Hyun Kim, eds. *Dreamscapes of Modernity: Sociotechnical Imaginaries and the Fabrication of Power* (Chicago: Chicago University Press, 2015).

Jenzen, Olu, and Sally R Munt, eds. *The Ashgate Research Companion to Paranormal Cultures* (Farnham: Ashgate, 2013).

Johnson, Nuala, 'Cast in Stone: Monuments, Geography and Nationalism', *Environment and Planning D: Society and Space*, 13 (1995), 51–65.

Johnson-Freese, Joan, *Heavenly Ambitions: America's Quest to Dominate Space* (Philadelphia: University of Philadelphia Press, 2009).

Johnson-Freese, Joan, and David Burbach, 'The Outer Space Treaty and the Weaponization of Space', *Bulletin of the Atomic Scientists*, 75 (2019), 137–141.

Jones, Dafydd, ed. *Dada Culture – Critical Texts on the Avant-Garde* (New York: Rodopi).

Jones, Dudley, and Tony Watkins, eds. *A Necessary Fantasy? The Heroic Figure in Children's Popular Culture* (London: Garland, 2000).

Jones, Rhys, 'Relocating Nationalism: On the Geographies of Reproducing Nations', *Transactions of the Institute of British Geographers*, 33 (2008), 319–334.

Jung, Carl, *Flying Saucers – A Modern Myth of Things Seen in the Skies* (London: Routledge & Kegan Paul, 1959).

Kearnes, Matthew, and Thom van Dooren, 'Rethinking the Final Frontier: Cosmo-Logics and an Ethic of Interstellar Flourishing', *Geohumanities*, 3 (2017), 178–197.

Keep, Christopher, 'H G Wells and the End of the Body', *Victorian Review*, 23 (1997), 232–243.

Kessler, Elizabeth, *Picturing the Cosmos – Hubble Space Telescope Images and the Astronomical Sublime* (London: University of Minnesota Press, 2012).

Kilgore, De Witt Douglas, *Astrofuturism: Science, Race and Visions of Utopia in Space* (Philadelphia: University of Pennsylvania Press, 2003).

Kirby, David A, *Lab Coats in Hollywood – Science, Scientists and Cinema* (Cambridge, MA: MIT Press, 2010).

Kitchin, Rob, and James Kneale, eds. *Lost in Space – Geographies of Science Fiction* (Trowbridge: Cromwell, 2002).

_____'Science Fiction or Future Fact? Exploring Imaginative Geographies of the New Millennium', *Progress in Human Geography*, 25 (2001), 19–35.

Klein, John J, 'Space Warfare: A Maritime-Inspired Space Strategy', *Astropolitics: The International Journal of Space Politics & Policy*, 2 (2004), 33–61.

Klinger, Julie, 'Environmental Geopolitics and Outer Space', *Geopolitics* (2019).

_____*Rare Earth Frontiers – From Terrestrial Subsoils to Lunar Landscapes* (Ithaca: Cornell University Press, 2017).

Krige, John, 'Atoms for Peace, Scientific Internationalism and Scientific Intelligence', *Osiris*, 21 (2006), 161–181.

Krige, John, and Arturo Russo, *A History of the European Space Agency, 1958–1987, Volume 1 – The Story of ESRO and ELDO, 1958–1973* (Noordwijk: ESA, 2000).

Lane, K Maria D, 'Geographers of Mars – Cartographic Inscription and Exploration Narrative in Late Victorian Representations of the Red Planet', *Isis*, 96 (2005), 477–506.

_____*Geographies of Mars: Seeing and Knowing the Red Planet* (Chicago: University of Chicago Press, 2011).

_____'Mapping the Mars Canal Mania: Cartographic Projection and the Creation of a Popular Icon', *Imago Mundi*, 58 (2006), 198–211.

Latour, Bruno, *Science in Action: How to Follow Scientists and Engineers Through Society* (Cambridge, Mass: Harvard University Press, 1987).

Launius, Roger D, *Apollo's Legacy – Perspectives on the Moon Landings* (Washington DC: Smithsonian Books, 2019).

Legett, Don, and Charlotte Sleigh, eds. *Scientific Governance in Britain, 1914–79* (Manchester: Manchester University Press, 2016).

Levine, George, *Darwin and the Novelists: Patterns of Science in Victorian Fiction* (Chicago: University of Chicago Press, 1991).

Livingstone, David N, *Putting Science in Its Place – Geographies of Scientific Knowledge* (Chicago: University of Chicago Press, 2003).

Livingstone, David N, and Charles W J Withers, eds. *Geography and Revolution* (Chicago: University of Chicago Press, 2010).

Macauley, William, 'Crafting the Future: Envisioning Space Exploration in Post-War Britain', *History and Technology*, 28 (2012), 281–309.

MacDonald, Fraser, 'Anti-Astropolitik: Outer Space and the Orbit of Geography', *Progress in Human Geography*, 31 (2007), 592–615.

————— *Escape from Earth - A Secret History of the Space Rocket* – (London: Profile, 2019).

————— 'Geopolitics and "the Vision Thing": Regarding Britain and America's First Nuclear Missile', *Transactions of the Institute of British Geographers*, 31 (2006), 53–71.

————— 'Space and the Atom: On the Popular Geopolitics of Cold War Rocketry', *Geopolitics*, 13 (2008), 611–634.

MacDonald, Fraser, and Charles W J Withers, eds. *Geography, Technology and Instruments of Exploration* (Farnham: Ashgate, 2015).

Maclaren, Andrew, 'Geopolitical Imaginaries of the Space Shuttle Mission Patches', *Geopolitics* (2019).

Mahony, Martin, 'Historical Geographies of the Future: Airships and the Making of Imperial Atmospheres', *Annals of the Association of American Geographers*, 109 (2019), 1279–1299.

Massie, Harrie, and M O Robins, *History of British Space Science* (Cambridge: Cambridge University Press, 1986).

Mawhinney, Rory, 'Astronomical Fieldwork and the Spaces of Relativity: The Historical Geographies of the 1919 British Eclipse Expeditions to Príncipe and Brazil', *Historical Geography*, 46 (2018), 203–238.

McAleer, Neil, *Odyssey: The Authorised Biography of Arthur C Clarke* (London: Victor Gollancz, 1992).

McCarthy, Patrick A, Charles Elkins, and Martin H Greenberg, eds. *The Legacy of Olaf Stapledon – Critical Essays and an Unpublished Manuscript* (London: Greenwood Press, 1989).

McCormack, Derek P, 'Geographies for Moving Bodies: Thinking, Dancing, Spaces', *Geography Compass*, 2 (2008), 1822–1836.

McCray, W Patrick, 'Amateur Scientists, the International Geophysical Year, and the Ambitions of Fred Whipple', *Isis*, 97 (2006), 634–658.

McCurdy, Howard E, *Space and the American Imagination* (Washington DC: Smithsonian Institution Press, 1997).

Merriman, Peter, and Rhys Jones, 'Nations, Materialities and Affects', *Progress in Human Geography*, 41 (2017), 600–617.

Messeri, Lisa, *Placing Outer Space – An Earthly Ethnography of Other Worlds* (London: Duke University Press, 2016).

Messeri, Lisa, and Janet Vertesi, 'The Greatest Missions Never Flown – Anticipatory Discourse and the "Projectory" in Technological Communities', *Technology and Culture*, 56 (2015), 54–85.

Miller, Ryder W, ed. *From Narnia to a Space Odyssey – The War of Ideas between Arthur C Clarke and C S Lewis* (New York: ibooks, 2003).

Millitz, Elisabeth, and Carolin Schurr, 'Affective Nationalism: Banalities of Belonging in Azerbaijan', *Political Geography*, 54 (2016), 54–63.

Montgomery, John, *The Fifties* (Leicester: Blackfriars Press, 1965).

Morton, Peter, *Fire Across the Desert – Woomera and the Anglo-Australian Joint Project, 1946–1980* (AGPS: Canberra, 1989).

Murray, Andy, *Into the Unknown – The Fantastic Life of Nigel Kneale* (London: Headpress, 2017).

Mussell, James, *Science, Time and Space in the Late Nineteenth-Century Periodical Press: Movable Types* (Aldershot: Ashgate Press, 2007).

Naylor, Simon, 'Introduction: Historical Geographies of Science – Places', *Contexts, Cartographies, The British Journal for the History of Science*, 38 (2005), 1–12.

Naylor, Simon, and James Ryan, eds. *New Spaces of Exploration – Geographies of Discovery in the Twentieth Century* (London: I B Tauris, 2010).

Neufeld, Michael, 'Weimar Culture and Futuristic Technology: The Rocketry and Spaceflight Fad in Germany, 1923–1933', *Technology and Culture*, 31 (1990), 725–752.

Nicholls, Peter, ed. *The Encyclopedia of Science Fiction* (London: Granada, 1981).

O'Connor, Ralph, 'From the Epic of Earth History to the Evolutionary Epic in Nineteenth-Century Britain', *Journal of Victorian Culture*, 14 (2009), 207–223.

Olson, Valerie, 'Political Ecology in the Extreme: Asteroid Activism and the Making of an Environmental Solar System', *Anthropological Quarterly*, 85 (2012), 1027–1044.

Parker, Martin, and David Bell, eds. *Space Travel and Culture: From Apollo to Space Tourism* (Oxford: Blackwell, 2009).

Parkinson, Bob, ed. *Interplanetary – A History of the British Interplanetary Society* (London: British Interplanetary Society, 2008).

Pearson, Alastair, and Others, 'Cartographic Ideals and Geopolitical Realities: International Maps of the World from the 1890s to the Present', *The Canadian Geographer*, 50 (2006), 149–76.

Phillips, Richard, *Mapping Men and Empire: A Geography of Adventure* (London: Routledge, 1997).

Poole, Robert, 'The Challenge of the Spaceship: Arthur C Clarke and the History of the Future, 1930–1970', *History and Technology*, 28 (2012), 255–280.

Redfield, Peter, eds. *Space in the Tropics – From Convicts to Rockets in French Guiana* (Berkeley: University of California Press, 2000).

_____'The Half-Life of Empire in Outer Space', *Social Studies of Science*, 32 (2002), 791–825.

Ribas, Ignasi, and Others, 'A Candidate Super-Earth Planet Orbiting Near the Snow Line of Barnard's star', *Nature*, 563 (2018), 365–368.

Roberts, Adam, *The History of Science Fiction* (New York: Macmillan, 2005).

Robertson, Francis, 'Science and Fiction: James Nasmyth's Photographic Images of the Moon', *Victorian Studies*, 48 (2006), 595–623.

Rodriguez, Mariano M, 'From Stapledon's Star Maker to Cicero's Dream of Scipio: The Visionary Cosmic Voyage as a Speculative Genre', *Foundation – The Review of Science Fiction*, 118 (2014), 45–58.

Rolinson, Dave, and Nick Cooper, '"Bring Something Back" – the Strange Career of Professor Bernard Quatermass', *Journal of Popular Film and Television*, 30 (2002), 158–165.

Sage, Daniel, 'Framing Space: A Popular Geopolitics of American Manifest Destiny in Outer Space', *Geopolitics*, 13 (2008), 27–53.

_____ *How Outer Space Made America: Geography, Organization and the Cosmic Sublime* (Farnham: Ashgate, 2014).

Said, Edward, *Orientalism* (London: Routledge, 1978).

Salin, Patrick, 'Privatization and Militarization in the Space Business Environment', *Space Policy*, 17 (2001), 19–26.

Sammler, Katherine, and Casey Lynch, 'Spaceport America: Contested Offworld Access and the Everyman Astronaut', *Geopolitics* (2019).

Saunders, Angharad, 'Literary Geography: Reforging the Connections', *Progress in Human Geography*, 34 (2010), 436–452.

Secord, Anne, 'Science in the Pub: Artisan Botanists in Early Nineteenth-Century Lancashire', *History of Science*, 32 (1994), 269–314.

Secord, James, 'How Scientific Conversation Became Shop Talk', *Transactions of the Royal Historical Society*, 17 (2007), 129–156.

Seedhouse, Eric, *Tim Peake and Britain's Road to Space* (Chichester: Springer Praxis, 2017).

Sharman, Helen, and Christopher Priest, *Seize the Moment – The Autobiography of Helen Sharman* (London: Victor Gollancz, 1993).

Sharp, Joanne, 'Hegemony, Popular Culture and Geopolitics: The Reader's Digest and the Construction of Danger', *Political Geography*, 15 (1996), 557–570.

Shaw, Philip, *The Sublime* (London: Routledge, 2006).

Shayler, David, and Ian Moule, *Women in Space – Following Valentina* (Berlin: Springer Praxis, 2005).

Sheehan, William, *The Immortal Fire Within – The Life and Work of Edward Emerson Barnard* (Cambridge: Cambridge University Press, 1995).

Shelton, Robert, 'The Mars-Begotten Men of Olaf Stapledon and H G Wells', *Science Fiction Studies*, 11 (1984), 1–14.

Sherborne, Michael, *H G Wells – Another Kind of Life* (London: Peter Owen, 2010).

Siddiqi, Asif, *Challenge to Apollo: The Soviet Union and the Space Race, 1945–1974* (Washington DC: NASA, 2000).

———*The Red Rockets' Glare – Spaceflight and the Soviet Imagination* (Cambridge: Cambridge University Press, 2010).

Smith, David C, ed. *The Correspondence of H G Wells, Volume 1, 1880–1903* (London: Pickering and Chatto, 1998).

Snowdon, Neil, ed. *We Are the Martians – The Legacy of Nigel Kneale* (Hornsea: Electric Dreamhouse, 2017).

Spiller, James, *Frontiers for the American Century: Outer Space, Antarctica and Cold War Nationalism* (New York: Palgrave Macmillan, 2015).

Spufford, Francis, *Backroom Boys – The Secret Return of the British Boffin* (London: Faber and Faber, 2003).

Squire, Rachel, and Klaus Dodds, 'Introduction to the Special Issue: Subterranean Geopolitics', *Geopolitics* (2019).

Stableford, Brian, *Scientific Romance in Britain, 1890–1950* (London: Fourth Estate, 1985).

Staeheli, Lynn A, Eleonore Kofman, and Linda J Peake, eds. *Mapping Women, Making Politics – Feminist Perspectives on Political Geography* (London: Routledge, 2004).

Stallabrass, Julian, *High Art Lite: British Art in the 1990s* (London: Verso, 2001).

Stewart, Larry, 'Other Centres of Calculation, or, Where the Royal Society Didn't Count: Commerce, Coffee-Houses and Natural Philosophy in Early Modern London', *The British Journal for the History of Science*, 32 (1999), 133–153.

Stillman, Danny B, and Thomas C Reed, *The Nuclear Express: A Political History of the Bomb and Its Proliferation* (Minneapolis: Zenith Press, 2010).

Stratton, Jon, *Britpop and the English Music Tradition* (London: Routledge, 2010).

Suvin, Darko, 'On the Poetics of the Science Fiction Genre', *College English*, 34 (1972), 372–382.

Sykes, Oliver, and others, 'A City Profile of Liverpool', *Cities*, 35 (2013), 299–318.

Uglow, Jenny, *The Lunar Men – The Friends Who Made the Future, 1780–1810* (London: Faber and Faber, 2002).

Wallace, Isabelle Loring, 'Technology and the Landscape: Turner, Pfeiffer and Eliasson after the Deluge', *Visual Culture in Britain*, 12 (2011), 57–75.

Wang, Sheng-Chih, 'The Making of New "Space": Cases of Transatlantic Astropolitics', *Geopolitics*, 14 (2009), 433–461.

Ward, Stuart, ed. *British Culture and the End of Empire* (Manchester: Manchester University Press, 2001).

Warf, Barney, 'Geopolitics of the Satellite Industry', *Tijdschrift voor Economische en Sociale Geografie*, 98 (2007), 385–397.

Warren, Saskia, 'Audiencing James Turell's Skyspace: Encounters between Art and Audience at Yorkshire Sculpture Park', *cultural geographies*, 20 (2012), 83–102.

Weight, Richard, *Patriots – National Identity in Britain 1940–2000* (London: Macmillan, 2002).

Wells, Herbert George, *Experiment in Autobiography. Discoveries and Conclusions of a Very Ordinary Brain (Since 1866)* (New York: J B Lippincott, 1967 [1934])

Williams, Alison, 'Beyond the Sovereign Realm: The Geopolitics and Power Relations in and of Outer Space', *Geopolitics*, 15 (2010), 785–793.

Winter, Frank, *Prelude to the Space Age – The Rocket Societies: 1924–1940* (Washington DC: Smithsonian Institution Press, 1983).

Wolfe, Gary, ed. *Science Fiction Dialogues* (Chicago: Academy Chicago, 1982).

Wolfe, Tom, *The Right Stuff* (New York: Bantam, 1979).

Science-fiction novels

Clarke, Arthur C, *Islands in the Sky* (Harmondsworth: Puffin, 1972 [1954]).

Clarke, Arthur C, *The Sands of Mars* (London: Sidgwick and Jackson, 1976 [1951]).

Clarke, Arthur C, *Earthlight* (London: Pan, 1963 [1955]).

Clarke, Arthur C, *Prelude to Space* (London: Sidgwick and Jackson Ltd., 1953).

Stapledon, Olaf, *Last and First Men* (Harmondsworth: Penguin, 1987 [1930]).

Stapledon, Olaf, *Star Maker* (London: Magnum, 1979 [1937]).

Wells, H G, *The War of the Worlds* (London: Penguin, 2005 [1898]).

Wells, H G, *The First Men in the Moon* (London: Penguin, 2005 [1901]).

Archives

BBC Written Archives Centre, Caversham Park, Reading.

British Film Institute National Archive, Fitzrovia, London.

British Interplanetary Society Archive, Vauxhall, London.

National Air and Space Museum Archive, Smithsonian Institution, Steven F Udvar-Hazy Center, Chantilly, Virginia, USA.

Special Collections and Archives, University of Liverpool.

The National Archives, Kew, London.

The Times Digital Archive

Index

Note: *Italicized* pages refer to figures.

Adams, Ansel 170
Adrift in the Stratosphere 44
Amazing Stories 48
American Interplanetary Society 36, 49
Americanism, in British outer space
   100–105
American Manifest Destiny 90
American Rocket Society 92;
   Astronautics 50; *Bulletin* 50
Andrews, E. D. G. 97
Anglo-Mysore Wars 36
Antequera Ltd 139
Anti-Ballistic Missile Treaty
   (1972–2002) 163
Apollo 38, 61, 106, 111, 116, 136, 138,
   140, 145
Apollonian gaze 74, 151
Apollo-Soyuz test project (1975) 135
Appleton Laboratory 117
Ariane rockets 91, 104
Ariel satellite 9, 101–103, 138
Aristotle 4
Armchair Science 44
Army Air Corps 152
Arnold, Kenneth 69
Asimov, Isaac 77
Askham, Colin 45
Astounding 48
'Astris' missile 103
Astrofuturism 61, 67, 78, 99
Astronautica Acta 95
astronautics 7, 37, 40–45, 47–51, 54–56,
   72, 73, 77, 97, 104, 127, 129, 161
Astronautics 50
astrophysics 2, 7, 13, 14, 17, 25, 26,
   105–106, 113
atomic power 101, 115–116, 124

banal nationalism 139–145
Barnard, Edward Emerson 119
Barnard's Star 9, 111, 117–121, 130
Beagle 2 Mars lander 9, 10, 135, 138,
   145–150
Beresford, John Davys 15
Bergson, Henri 17
Berlin Dada art movement (1920) 94
biographical cultures of science 15–18
Birch, Andrew: 'Young British Artists'
   149, *149*
BIS *see* British Interplanetary
   Society (BIS)
BIS Journal (*JBIS*) 9, 41, 44, 49, 50, 73,
   79, 80, 93, 94, 114, 115; 'Interstellar
   Studies' 126, 128, 129; space-ship
   51–55, *53*
Black Arrow 104
Blair, Tony 148
'Blue Streak' missile 103–105
Bond, Alan 118, 120, 123, 124,
   127–129
Bonestell, Chesley 90, 170
Bramhill, H. 52
Braun, Wernher von 61, 94
British Aircraft Corporation 117
British Astronomical Association 79
British Interplanetary Society (BIS)
   1–3, 78, 88, 91–92, 104, 106, 111,
   114, 161; astronautics 40–45; 'BIS
   Space-Ship' 7–8; Commonwealth
   Spaceflight Symposium, London,
   August 1959 97; interstellar explora-
   tion 9, 111–131; in Liverpool 44–46,
   48, 52; in London 46–47, 52; networking
   47–51; outer space, synthesising
   36–56; *Spaceflight* 8, 73–75, 97,
   100, 102, 164; space-ship 51–55, *53*;
   Technical Committee 52

British space programme 88–106;
 Americanism in 100–102;
 Commonwealth space programme
 96–100; critical astropolitics and
 outer space 89–91; Europeanism in
 100–105; internationalism in outer
 space 91–96, *95, 96*; limits of moder-
 nity 105–106; New Elizabethanism
 96–100, 105
British UFO Research Association
 (BUFORA) 8, 70–72
Brooks, Gordon 140
Brügel, Werner 49
BUFORA *see* British UFO Research
 Association (BUFORA)
BUFORA Journal 70
Bulletin 50
Burgess, Eric 113–114
Burke, Edmund 169
Burroughs, Edger Rice 12
Bush, George W. 89

Calder, Nigel: *Spaceships of the Mind*
 127, 128
California, 1930s' experiments of
 rocketry pioneers in 3
Campbell, John 48
Cartier, Rudolph 63
Centerbladet 50
Centre Spatial Guyanais 91
Challenger disaster (1986) 139
Christian Science Monitor 113
Cicero: *Dream of Scipio* 24
Clare, Roy 147
Clarke, Arthur C. 1–4, 8, 9, 46, 48, 52,
 61, 62, 72, 76–83, 99, 104, 105, 122,
 128, 161; *Earthlight* 78–80; *Islands
 in the Sky* 78–79, 81–82; *Prelude to
 Space* 81, 98; *Sands of Mars, The* 78,
 79; space trilogy, language of inter-
 planetarism in 77–82
class 22, 27, 29; middle 17, 19, 41, 77; social
 6, 16, 17, 91; working 17, 18, 41, 137
Cleator, Philip Ellaby 7, 37–46,
 48–52, 55, 73, 76; 'Possibilities of
 Interplanetary Travel, The' 42;
 *Rockets Through Space, Or, The
 Dawn of Interplanetary Travel* 42, 49,
 51, 55; *see also* British Interplanetary
 Society (BIS)
Cleaver, Arthur 'Val' 44, 46, 47, 76, 94,
 103, 104; 'Interplanetary Project,
 The' 92

Cleveland Rocket Society: *Space* 50
Clinton, Bill 145
cognitive estrangement 14, 31, 161
Cold War 3, 18, 39, 66, 68–70, 88, 89,
 91–93, 98, 99, 101, 103, 105, 115–117,
 135–137, 141, 156
colonialism 90, 91
Commonwealth space programme
 96–100
Congreve, Sir William 36
Constable, John 169
Constitution for the IAF 93
'Coralie' missile 103
'Corporal' missile 138
Corrêa, Henrique Alvim 27, *28*
Cosgrove, Denis 1, 2, 4, 5, 161
cosmic ascent 15, 16, 30, 36, 170
cosmic transcendence 36, 55
cosmos 2, 3, 7, 9, 10, 13, 22, 24–26, 30,
 48, 54, 56, 62, 112, 117, 127, 130, 171,
 173–175
critical astropolitics, and outer space
 89–91
cyberpunk 14

Daily Express, The 42
Daily News, The 27
Daily Sketch, The 46
Dan Dare 6, 62, 70
DARPA, '100 Year Starship' 129
Darwin, Charles 19, 146, 150
Das Neue Fahrzeug 50
Defence Arctic Strategy 2018 (UK) 167
Derrida, Jacques 67
Diamond, Anne 140
Die Rakete 50
Dillon, Bill 122, *122*
Doctor Who 62–63
Drake, Frank 113
Drayson, Lord 150, 151
Dyson, Freeman 115, 127
Dyson Sphere 127

Eddington, Arthur 17
Edwards, J. H. 52
Einstein, Albert 25
ELDO *see* European Launcher
 Development Organisation (ELDO)
Eliasson, Olafur: 'Weather Project,
 The' 170
Elizabeth I 98
Elizabeth II 91, 98
Emin, Tracey 147

Equatorial Launch Australia, Arnhem 163
ESA *see* European Space Agency (ESA)
Esnault-Pelterie, Robert 36
ESRO *see* European Space Research Organisation (ESRO)
ethnicity 6, 114
eugenics 13, 16, 20, 21, 27
European astroculture 3
European Economic Community 104
Europeanism, in British outer space 100–105
European Launcher Development Organisation (ELDO) 9, 103–105
European Space Agency (ESA) 91, 97, 103, 135, 146, 149–152, 164, 173; Astronaut Corps 151, 155; spaceport of 167
European Space Research Organisation (ESRO) 103, 104
European Union: Galileo project 164
EVAs *see* Extra-Vehicular Activities (EVAs)
Evening Journal 27
Everett, Cornelius J. 115
evolutionary theory 7, 13, 14, 19–21
evolutionism 7, 13, 16, 21, 25, 31, 61, 78, 82
Explosives Act of 1875 (UK) 40–41, 45
extra-terrestrial intelligence 68, 71, 112–118
Extra-Vehicular Activities (EVAs) 152, 153

Fabian Society 17
Fermi, Enrico 112, 116
Fermi's Paradox 9, 112, 129
Fetherstonhaugh, Sir Harry 16
First Hand 77
First International Peace Conference in Paris (1949) 94
First World War 17, 94
Flammarion, Camille 12
Flying Saucer from Mars 77
Flying Saucer Working Party 8, 69, 70
Frau im Mond (film) 39
French Aero Club 92
Fren, Richard 29

Gagarin, Yuri 137–138, 142
Galileo 2
Galton, Francis 21
gender 137, 138, 157; and identity 38; and mobility 38; and space exploration 6; stereotypes 114
generation ship 129

geology 23–24, 26, 116, 169
geopolitics: critical 89–90; definition of 89; and interstellar exploration 112–117; and space exploration 8–9
Gernsback, Hugo 48
Gesselschaft für Weltraumforschung (GfW) 92, 94, *95, 96*
GfW see Gesselschaft für Weltraumforschung (GfW)
Glavkosmos 139
Globe 28
Goddard, Robert 36, 40, 52, 94
Gorbachev, Mikhail 141
Grady, Monica 145
Graphic, The 28
Gravitation law 30
Griffith, George 15
Groupement Astronautique Français 92

Hammer studios 63
Hanson, M. K 52
Hardy, David 128
haunted culture 68, 72
Hauntology 67
Hawker Siddeley Dynamics 117
Heinlein, Robert 61, 77
helium-3, 123–124
Hello 155
Hevelius, Johannes 169
Highlands and Islands Enterprise 165
Hillary, Edmund 141
Hirst, Damien 10, 147, 148, 169
Horizon 126
Hovhannes, Alan: *Odysseus Symphony* 128
Hubbard, F/Lt 69–70
Hubble's Law 25
Humboldt, Alexander von: *Cosmos* 2, 4
Huxley, Thomas Henry 16, 20, 21

IAC *see* International Astronautical Congress (IAC)
IAF *see* International Astronautical Federation (IAF)
Icarus Interstellar 129
Ido–English Dictionary 50
Ido Magazine, The 50
Ido Society of Great Britain 50
Illustrated London News 31
Imperial Russia: scientific networks in 3
International Astronautical Congress (IAC) 94–95, *95, 96*
International Astronautical Federation (IAF) 92, 93, 95, 96, 105

International Geophysical Year
  (1957–8) 74, 101
International Map of the World 93
International Space Station (ISS)
  152–155, 173
interplanetary science, spatiality of
  55–56
interplanetary space 20–22
interstellar exploration 9, 111–131;
  Barnard's Star 117–121; extra-terrestrial intelligence and 112–117; geopolitics and 112–117; Project Daedalus:
  cultural and scientific impressions
  126–130, *130*; Solar System environment, reconceptualising 121–126, *122*
Interstellar Studies 9
Irish Home Rule 41
ISS *see* International Space Station (ISS)

James, Alex 148
James, William 17
Janser, A. 52
Jeanes, James 17
Jet Propulsion Laboratory 38–39
Jodrell Bank radio telescope 101
Johnson, Leslie 45, 48–50
Johnstone, Paul 74
Jung, Carl 68

Kant, Immanuel 169
Kármán, Theodore von 36, 40
Kelly, Scott 156
Kensington Central Library,
  London 71
Kepler, Johannes 2, 120–121;
  Somnium 14
Kneale, Nigel 8, 63–67, *63*, 83
Kopra, Tim 152
Kourou rocket range, French Guiana
  90–91, 103
Kubrick, Stanley 128

Lahav, Ofer 171
Lancashire Daily Post 29
landscape 4, 6, 12–14, 16, 22–24, 26, 31,
  40, 66, 79, 90, 98, 136, 162–170, 173
Lang, Fritz 39
Lasser, David 37
LaunchUK 165–166, *166*, 168
Lawton, Tony 118–120
Lewis, C. S. 6
Ley, Willy 46, 49, 50
literary geographies, and science fiction
  13–15

Liverpool Daily Post 31
Liverpool Echo 42
Lockheed Martin 165, 166
Loeser, Guenter 93
London Congress of the IAF 93, 94
London Standard 29
Lovell, Bernard 102
Low, A. M. 7, 41, 43–44, 51, 55
Lowell, Percival 19
Lucian: *Icaro-Menippus* 2

Mace, Major Timothy 140, 141
Mackinder, Halford 89
Mahan, Alfred T. 89
Malenchenko, Yuri 152
Malina, Frank 36, 38–40
Manchester Interplanetary Society 37
Manhattan Project 115
Martin, Tony 118, 120, 123, 124,
  128, 129
Matloff, Gregory L.: 'World Ships' 129
Mencken, Henry L. 42
Ministry of Defence (MoD, UK):
  Flying Saucer Working Party 69
Mir 135, 139–145, 152
Molton, P. 128–129
Montala Letro 50
Moonwatch 74
Moore, Patrick 5, 62, 81, 83; *Sky at
  Night, The* 72–77, *76*, 126, 128
Morning Post 28–29
Morris, William 17

Napoleonic Wars 36
NASA 101, 138, 145; Goddard
  Spaceflight Center 102; Jet
  Propulsion Laboratory 113; *Mars
  Pathfinder* missions 90; Pathfinder
  probe 146; Pioneer programme 125;
  Space Shuttle missions 153; Viking
  probes 112; Viking programme 146;
  '100 Year Starship' 129
Nasmyth, James 169
nationalism 90; banal 139–145; and
  cultures of outer space 135–138; and
  space exploration 135–157
National Maritime Museum, London
  147, 174
National Science and Media Museum,
  Bradford 155
National Space Centre (UK) 60n89,
  143–144
Nature 17, 55, 145
neo-classical astropolitics 89, 90

networking 47–51
New Elizabethanism 9, 96–100, 105
New Review, The 26
New York Post 42
Nuclear Decommissioning
    Authority 165
nuclear rocket propulsion 115
Nuclear Test Ban Treaties 115
Nuttall, Nick 140

Oberth, Hermann 36, 40, 52, 94
O'Neill, Gerard K. 127
OneWeb 164
Opel, Fritz van 52
Ortelius, Abraham 4
outer space: British internationalism
    in 91–96, *95, 96*; critical astropolitics
    and 89–91; culture, nationalism and
    135–138; domestication of 72–77,
    76; fiction of Wells and Stapledon
    12–32; geographies of 1–10, 37–40;
    in post-war Britain 61–83; synthe-
    sising 36–56; *see also* British space
    programme
Outer Space Treaty (1967) 102, 163
Ovenden, Michael 80

Paglen, Trevor 170
Paisley Rocketeers' Society 37
Pall Mall Gazette 27
Paris Congress of the IAF 92, 93
Parkinson, Bob 117, 121, 123–125
Paterson, Katie 10, 168–173; *All the
    Dead Stars* 171, *171*, 172; *Campo
    del Cielo* 172–173; *History of
    Darkness* 172
Peake, Tim 9, 10, 135, 138, 150–156, *151*,
    164, 174
Pearson's Magazine 26
Pendray, George Edward 36, 49
Perelman, Iakov 49
Picasso, Pablo: 'Dove of Peace' 94
Pillinger, Colin 10, 147, 148, 150; *Beagle
    2 Diaries* 146
Pioneer 11 125
Pioneer plaque 113, 114
Pirquet, Guido von 49
popular culture, in post-war Britain
    61–83; Clarke's space trilogy, lan-
    guage of interplanetarism in 77–82;
    outer space, domestication of 72–77,
    76; *Quatermass*, myths and moder-
    nity in 62–67, *63*; Unidentified Flying
    Object 62–67

Practical Mechanics 41
Project Daedalus 9, 111, 112, 117–131;
    cultural and scientific impres-
    sions 126–130, *130*; Solar System
    environment, reconceptualising
    121–126, *122*
Project Fishbowl 102
Project Grudge 69
Project Icarus 126, 129
Project Juno 139–142
Project Orion 115, 116, 124
Project Sign 69
'Prospero' satellite 104, 164
Ptolemy 4
Public Libraries Act of 1850 (UK) 26

Quatermass 8, 62, 75; myth and
    modernity in 62–67
Quatermass and the Pit 8, 62; myths and
    modernity in 66–67
Quatermass Experiment, The 8, 62;
    myths and modernity in 64–65, 66
Quatermass II 8, 62, *63*; myths and
    modernity in 65

race 6, riots (1958) 66–67
radio astronomy 6
RAF *see* Royal Air Force (RAF)
Raymond, N. E. Moore 42
Reade, William Winwood 17
Regan, Terry *130*
relativity, theory of 25
Robinson, Kim Stanley 90
Rocket Lab's Launch Complex 1,
    Mahia, New Zealand 163
Rocket Propulsion Establishment 117
Rockets Through Space 42
Romanticism 169
Rosny, J.-H. 37
Ross, H. E. 46, 52, 60n89, 154, 118
Rothko, Mark 169
Rowntree, Dave 148
Royal Aircraft Establishment 70
Royal Air Force (RAF) 69
Royal Naval College 154
Royal Navy 147
Royal Observatory 154
Royal Ordnance College 43
Russell, Bertrand 26, 30

Sagan, Carl 38, 113
Sagan, Linda Salzman 113, 114
Sanger, Eugene 52, 94
Schiaparelli, Giovanni 19

science fiction 12–32; biographical cultures of science 15–18; definition of 14; literary geographies and 13–15
Science Museum, London 155, 174
scientific romance 12, 14, 15, 18, 21, 27, 41
Scoops 48
Second World War 3, 8, 18, 27, 36, 37, 39, 40, 54, 58n43, 61, 65, 69, 83, 88, 92, 113, 115, 126, 147, 162
Seward, F. D. 129
Shackleton, Ernest 141
Shapley, Harlow 76
Sharman, Helen 9–10, 135, 138–145, *142*, *144*, 151, 152, 156
Shelley, Mary: *Frankenstein* 65
Shepherd, Leslie 92, 93; on generation ship 129; 'Interstellar Flight' 114; on spiritual satisfaction 114–115; starship concept 115–116
Sibelius, Jean 75; *Pelléas et Mélisande* 128
Sky at Night, The (BBC tv programme) 8, 72–77, 126, 128
Skylark sounding rocket 101
Smith, Clive 140
Smith, R. A. 52, 60n89, 122
socialism 17, 25, 136
Sokol spacesuit, 144, 155, 156
Southall, Ivan: *Rockets in the Desert* 99
South Kensington, London 7
Soyuz TM-12, 141, 142
Soyuz TMA-19M 152
Space 50
Space Age 3, 8, 38, 51, 61, 68, 72, 100, 104, 105, 111, 123, 139, 165, 169
space exploration 1–10, 88; European 3; fictional representations of 6; gender and 6; geopolitics and 8–9; nationalism and 135–157; as transcendental state 136; *see also individual entries*
Spaceflight 8, 73–75, 97, 100, 102, 164
Space Industry Act of 2018 164–165
Spaceport America, New Mexico 163, 167
spaceports 163–168
space race 3, 8, 88, 96
space science 1, 32, 38–42, 44, 71–73, 77–79, 83, 102, 103, 113, 117, 128, 139, 145, 150, 163, 164; geopolitics of 38; internationalism in 47–51
Sputnik 3, 73–74, 100, 115

Stapledon, Olaf 6–7, 12–15, 22–26, 36, 41, 45, 46, 82, 173; on cosmos 22, 24–25; 'Interplanetary Man' 116; on landscape 22–24, 26; *Last and First Men* 15, 18, 22–26, 30, 174; life-paths 15–18; readings of the Universe, Earth, and humanity 26–31; *Star Maker* 15, 22, 24–27, 30, 31, 172; on sublime 22–24, 26
'Starfish Prime' nuclear test 102
Steinitz, Otto 49
Stranger from Space 62
Strauss II, Johan: *Blue Danube, The* 128
Strong, James 118; *Flight to the Stars* 115
sublime 10, 13, 16, 22–24, 26, 30, 31, 61, 63, 66, 67, 83, 90, 115, 128, 145, 162, 163, 168–173, 175
Suvin, Darko 14

Tau Zero Foundation 129
TeleVista 43
Temple, William F. 48
Thatcher, Margaret 141
Them and the Thing 77
Thomas, Dylan 13
Thompson, Gordon 92, 97
Tikaram, Tanita 143
Times, The 101, 119, 140
Tizard, Sir Henry 69
Tomorrow's World 126
Traux, Robert 47
Truman, Harry 94
Tsiolkovskii, Konstantin 36, 39, 40
Turner, J. M. W. 169, 170
Turney, G. L. 69
Turrell, James: 'Skyspaces' 170
24 Ophiuchi 118
2001: A Space Odyssey 128
Tziolas, Andreas 129

Ufology 68–69
UFO *see* Unidentified Flying Object (UFO)
UKSA *see* UK Space Agency (UKSA)
UK Science Research Council 102
UK Space Agency (UKSA) 150, 151, 164, 165, 174
Ulam, Stanislaw M. 115
Unidentified Flying Object (UFO) 8, 62, 67–72, 77, 83, 161, 172
United States Air Force 115
Universal Science Circle, The 41

UN Outer Space Treaty (1967) 89
US–UK Mutual Defence Agreement
    (1958) 101

van de Kamp, Peter 118, 119
Verein für Raumschiffahrt (*VfR*, Society
    for Spaceflight) 36, 39, 48–50, 52; *Die
    Rakete* 50
Verne, Jules 5, 12, 81
VfR see Verein für Raumschiffahrt
    (*VfR*, Society for Spaceflight)
Voyager 1, 38, 113
Voyager probes 114
V–2 rocket 64, 93

Weedall, Norman 45
Weisman, Reid 153
Welles, Orson 27
Wells, H. G. 5–7, 12–15, 36, 41, 66,
    67, 78, 82, 145, 173; *First Men in the
    Moon, The* 15, 21, 22; on interplan-
    etary space 20–22; *Island of Doctor
    Moreau, The* 12; life-paths 15–18;

readings of the Universe, Earth, and
    humanity 26–31; scientific romances
    18–21; on society 20; *Time Machine,
    The* 12; *War of the Worlds, The* 15,
    19–22, 27–29, 65, 174
Whitstable Times 29
Wholey, Max W. 79
Willesden Chronicle, The 29
Willetts, David 151
Winkler, Johannes 36
Winter, Frank 42
Wireless World 79
Wolff, Heinz 139
Wonder Stories Quarterly 49
Woomera rocket range, South Australia
    9, 98–99, *100*, 167
Workers' Educational Association 17
Wright, Ian 145
Wright, Sydney Fowler 15

X Club, The 20

Yorkshire Post 42